Dimensions Math
Tests K

Published by Singapore Math Inc.

19535 SW 129th Avenue
Tualatin, OR 97062
www.singaporemath.com

Dimensions Math® Tests Kindergarten
ISBN 978-1-947226-46-3

First published 2019
Reprinted 2019, 2020 (twice)

Printed in China

Acknowledgments

Editing by the Singapore Math Inc. team.
Design and illustration by Cameron Wray with Carli Fronius.

Preface

Dimensions Math® Tests is a series of assessments to help teachers systematically evaluate student progress. The tests align with the content of Dimensions Math K–5 textbooks.

Dimensions Math Tests K uses pictorially engaging questions to test student ability to grasp key concepts through various methods including circling, matching, coloring, drawing, and writing numbers.

Dimensions Math Tests 1–5 have differentiated assessments. Tests consist of multiple-choice questions that assess comprehension of key concepts, and free response questions for students to demonstrate their problem-solving skills.

Test A focuses on key concepts and fundamental problem-solving skills.

Test B focuses on the application of analytical skills, thinking skills, and heuristics.

Contents KA

Contents KB

Chapter	Test	Page

Test 1

Chapter 1 Match, Sort, and Classify

(5 points each)

Circle the bowl on the right.

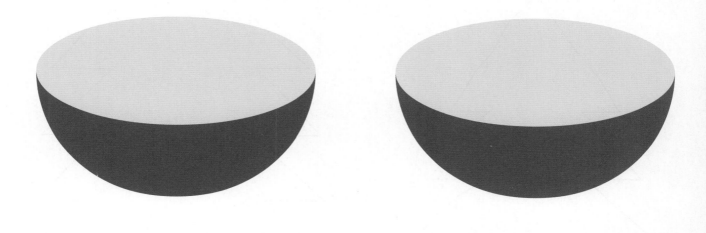

Circle the bird on the left.

Color the tree on the left.

Circle the flowers that are exactly the same.

Circle the things that are similar.

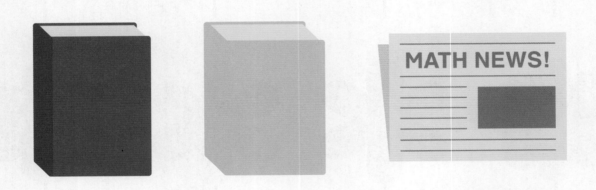

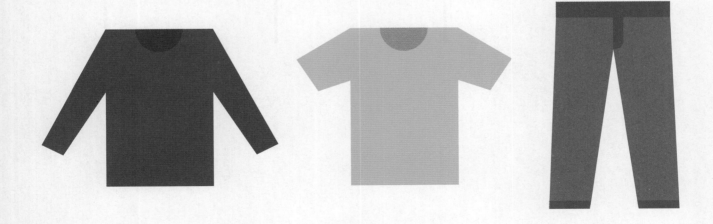

Cross out the one that is different.

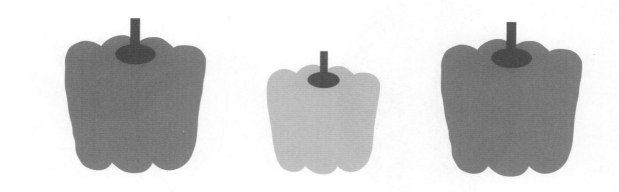

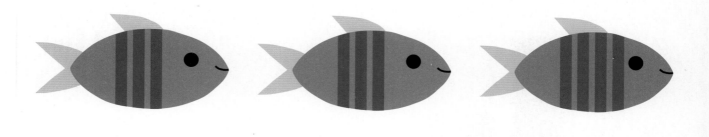

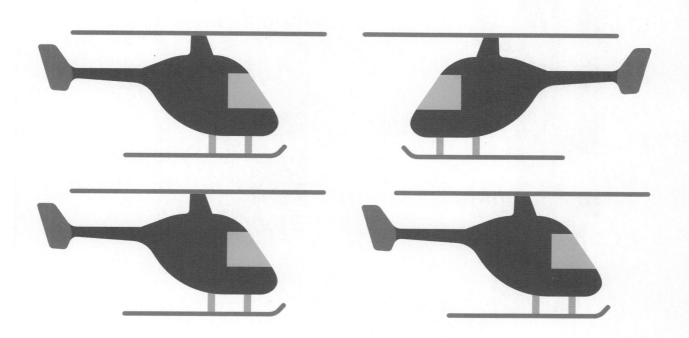

Circle the things that are rough.

Match objects that go together.

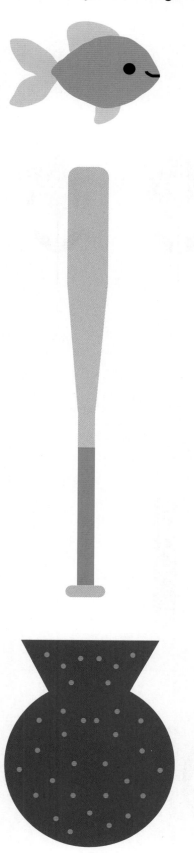

Cross out the thing that is not like the others.

Test 2

Chapter 2 Numbers to 5

(5 points each)

Circle the groups of 2.

Match.

Match.

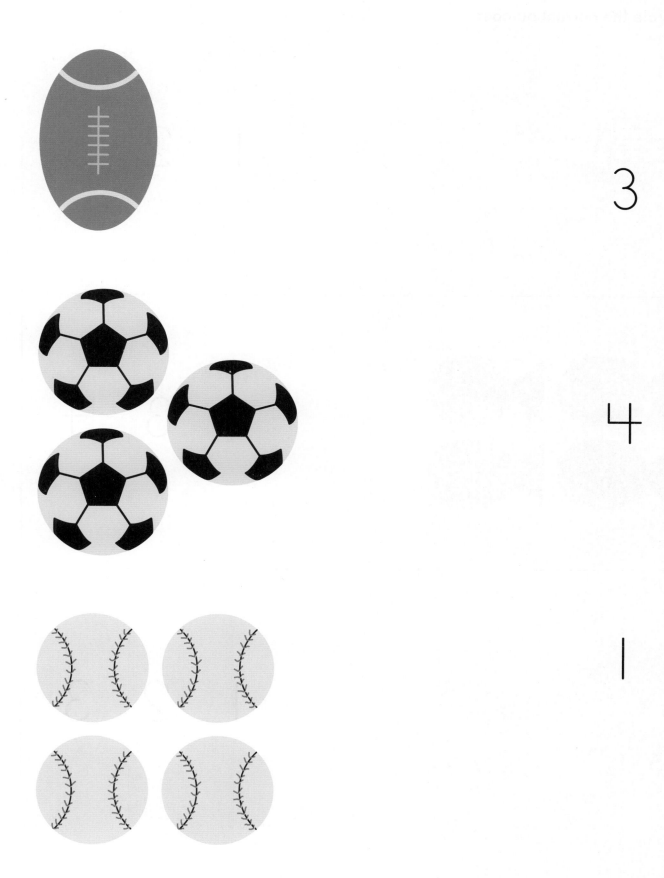

3

4

1

Count.
Circle the correct number.

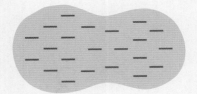

1   2   3

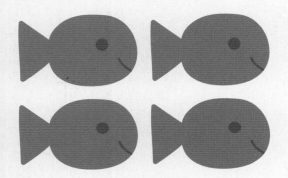

2   3   4

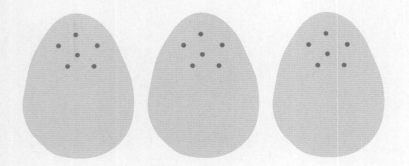

1   2   3

Circle the correct number of cats in each row.

2	
4	
5	
3	

Match.

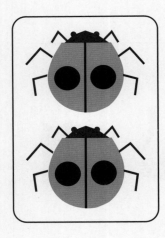

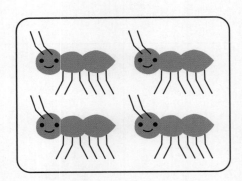

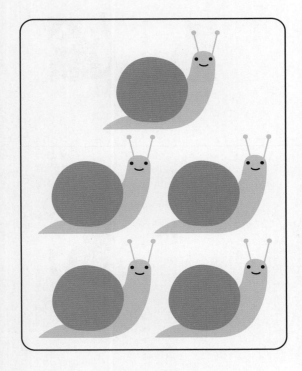

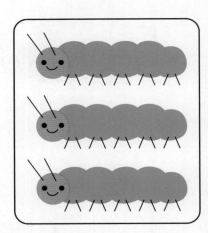

Test 3

Chapter 2 Numbers to 5

(5 points each)

Count and write the number.

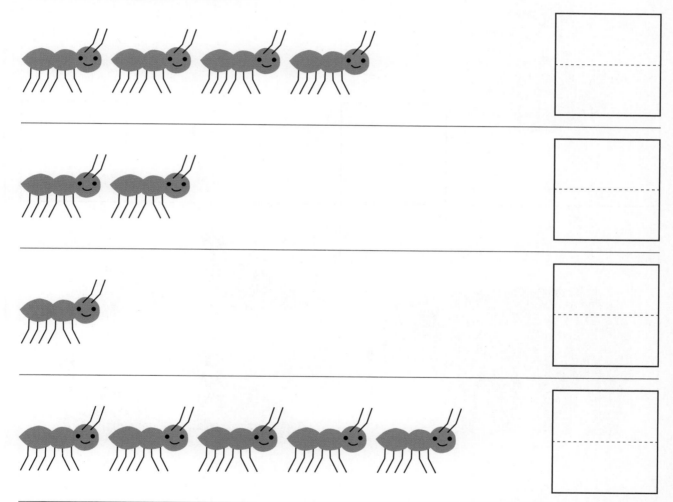

Count and write the number.

Count and draw the same number of dots in a different way.

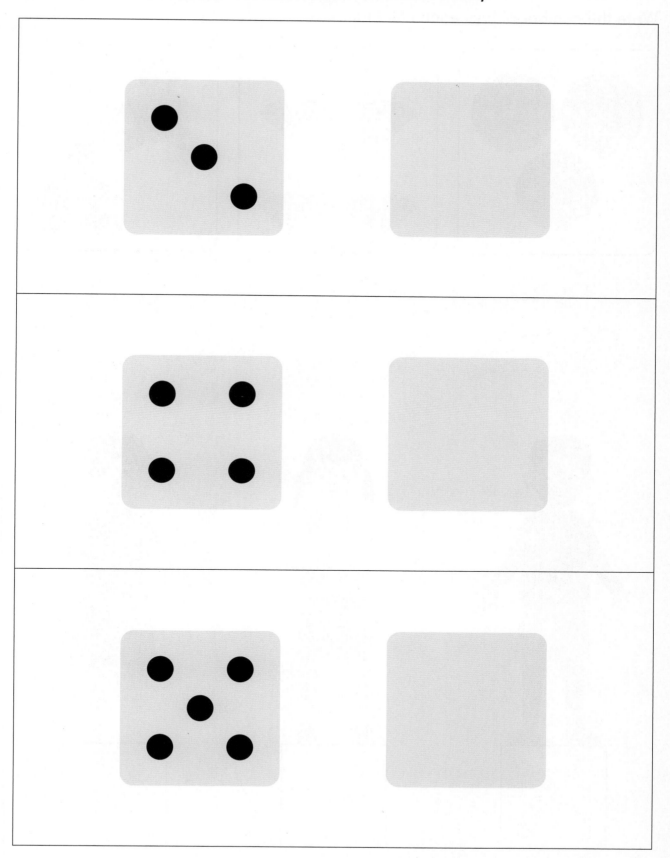

Trace a line to each child.
Write the number of toys each child has.

How many bears are there?
Circle the number.

5   3   4   1

Write the number to show how many.

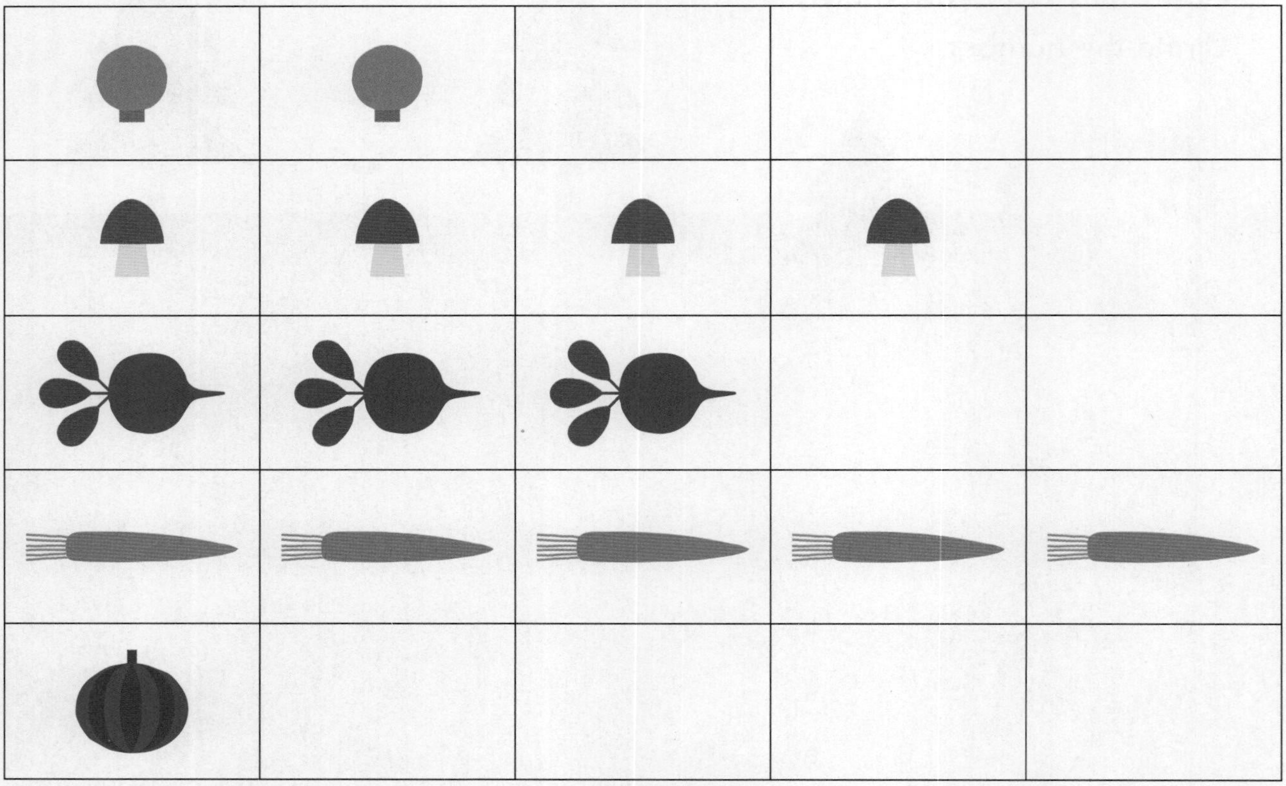

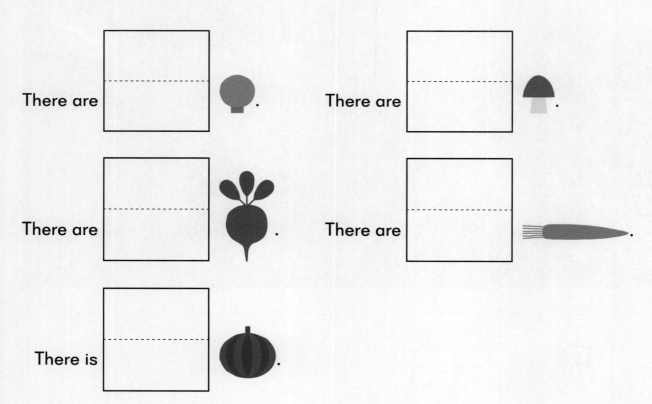

There are ____ 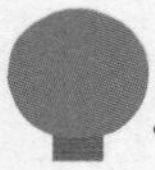.

There are ____ 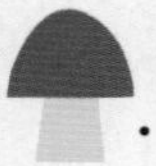.

There are ____ 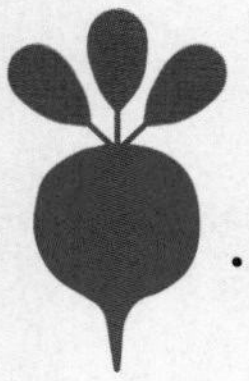.

There are ____ .

There is ____ 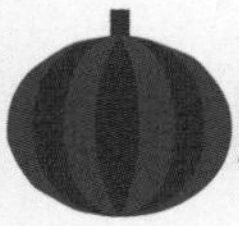.

Test 4

Chapter 3 Numbers to 10

(5 points each)

Circle the groups of 6.

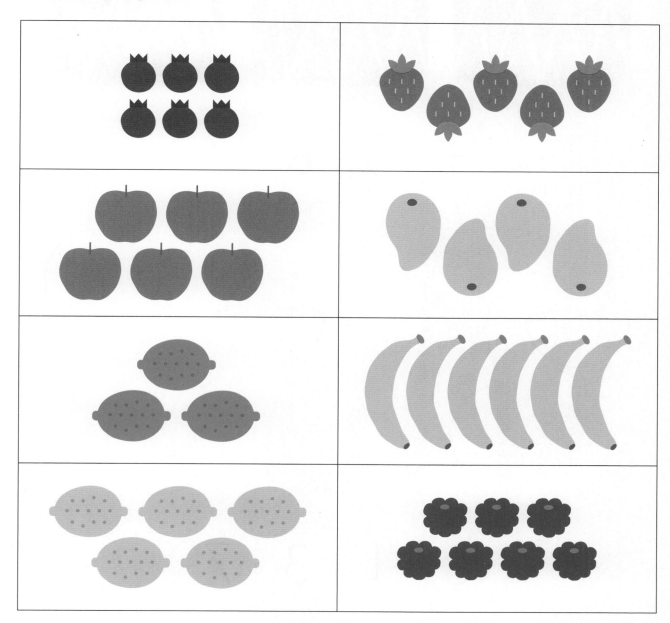

Count and circle the correct number.

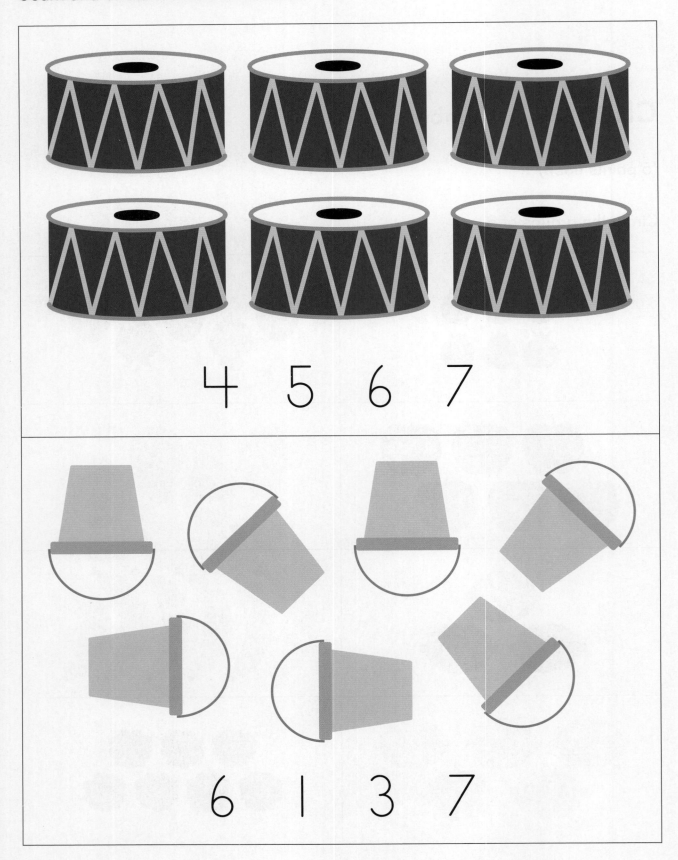

4   5   6   7

6   1   3   7

Circle the groups of 8.

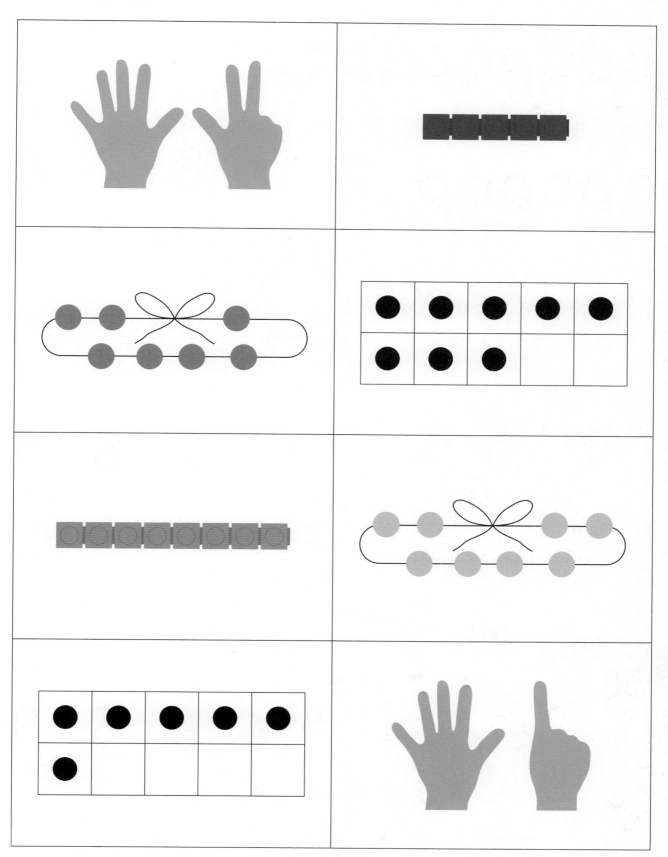

Count and circle the correct number.

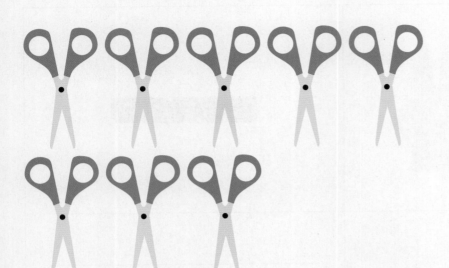

8   7   6

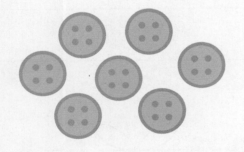

7   8   9

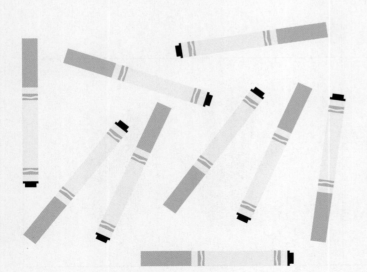

6   5   9

Write an X in the boxes with 10 items.

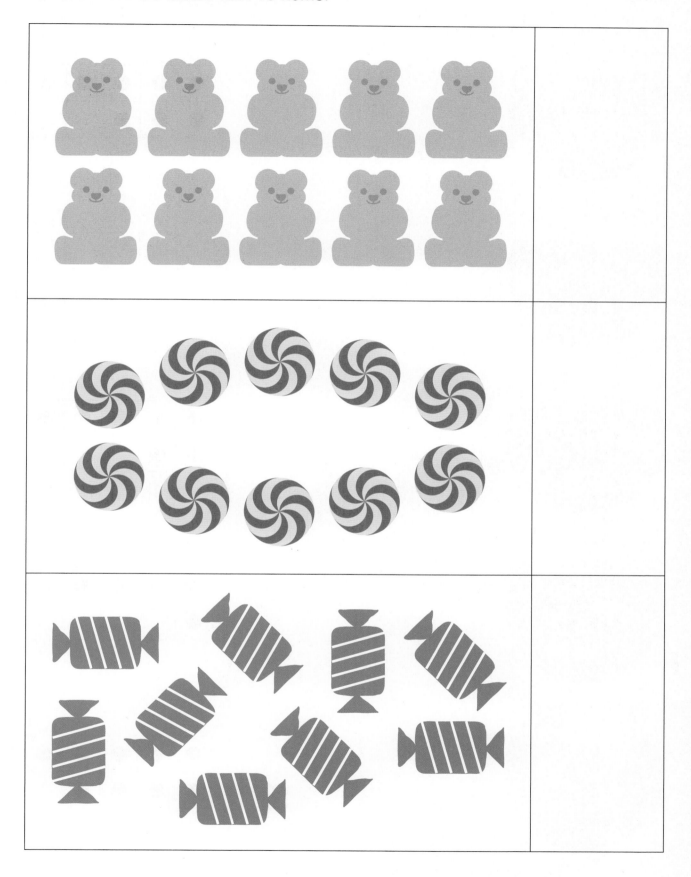

Match.

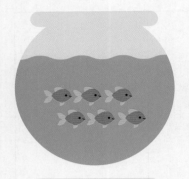

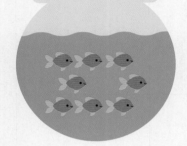

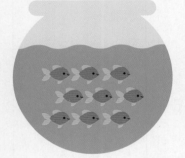

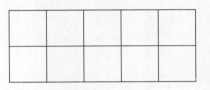

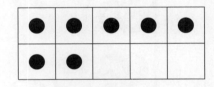

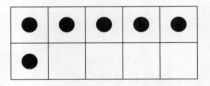

Count and color the correct number of boxes.

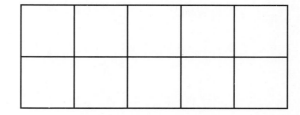

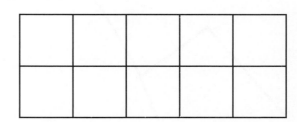

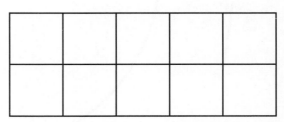

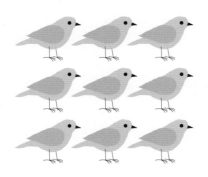

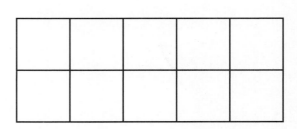

Draw 10 spots on the ladybug.

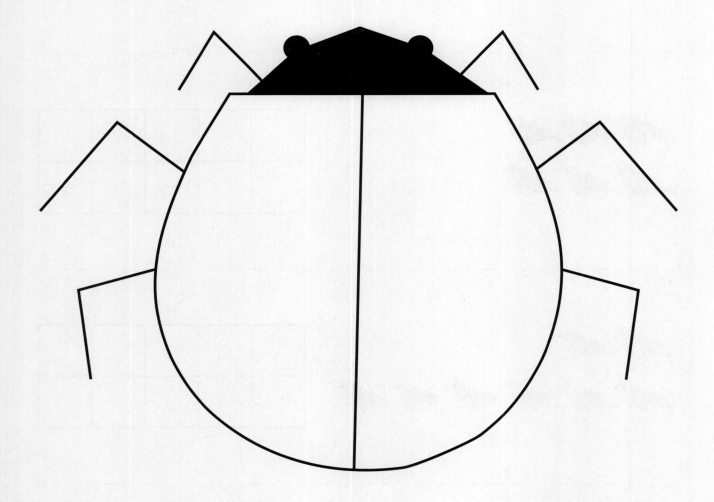

Chapter 3 Test 4

Name: _____   **Date:** _____

Test 5

Chapter 3 Numbers to 10

(5 points each)

Write the number in the box.

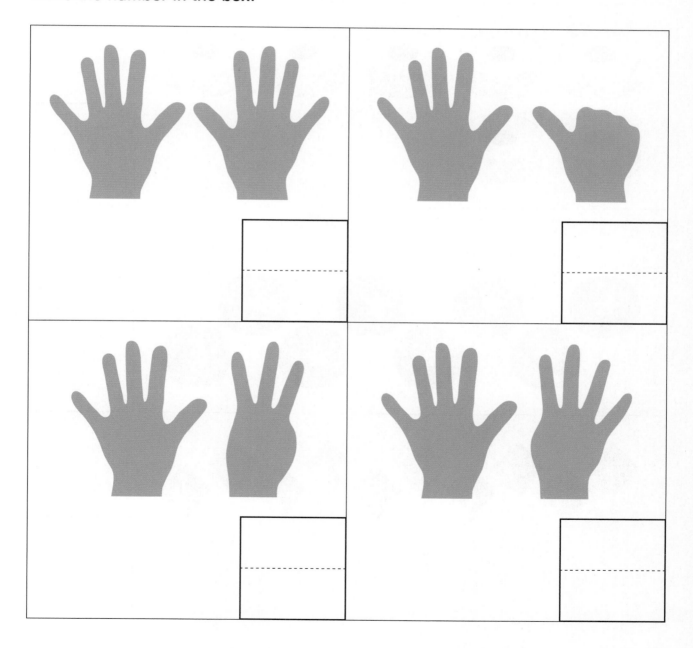

Count and write the number in the box.

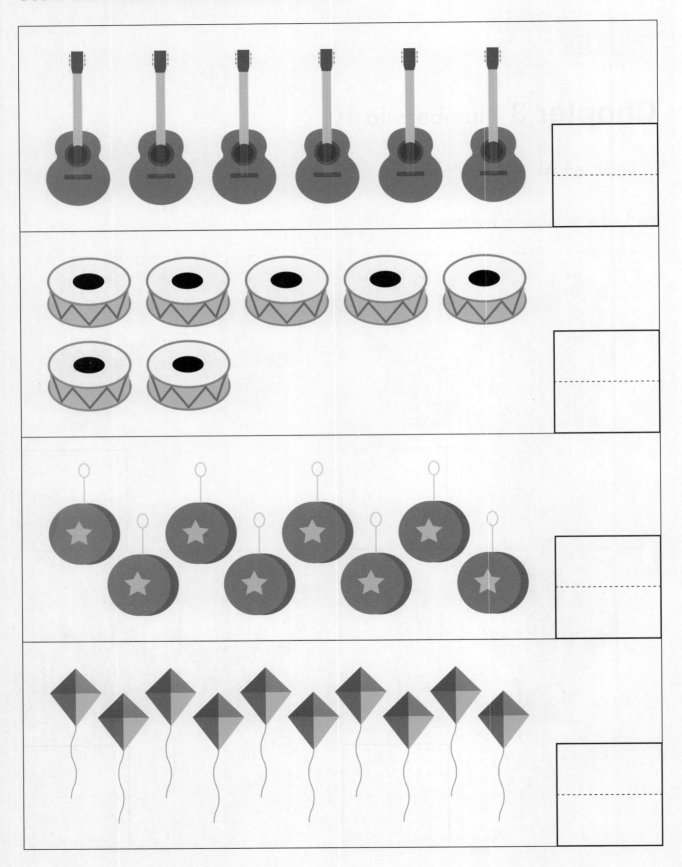

Count and write the number in the box.

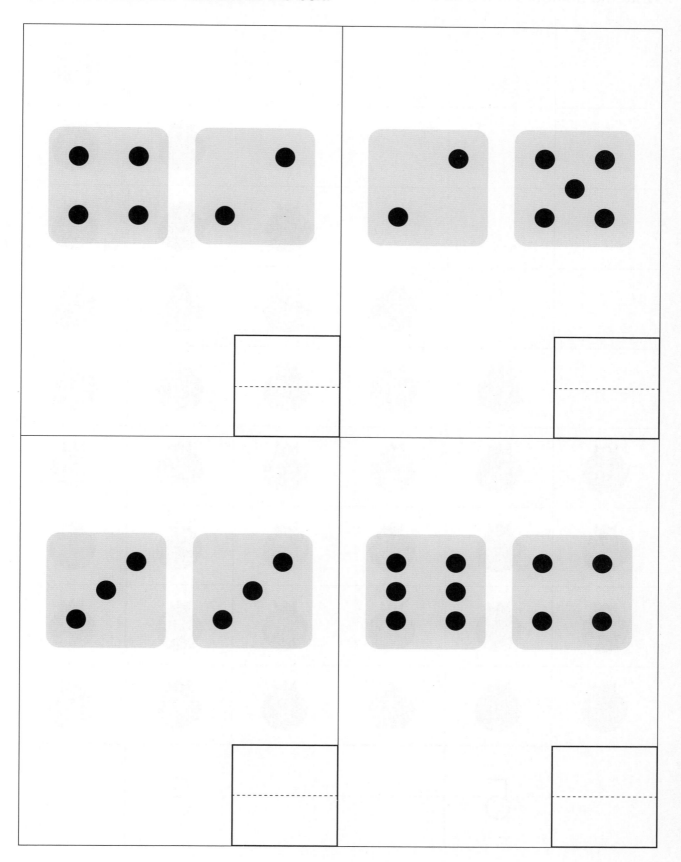

Write the numbers in the boxes.

					🮑
				🮑	🮑
			🮑	🮑	🮑
		🮑	🮑	🮑	🮑
	🮑	🮑	🮑	🮑	🮑
🮑	🮑	🮑	🮑	🮑	🮑
🮑	🮑	🮑	🮑	🮑	🮑
🮑	🮑	🮑	🮑	🮑	🮑
🮑	🮑	🮑	🮑	🮑	🮑
	5			8	

Write the missing numbers.

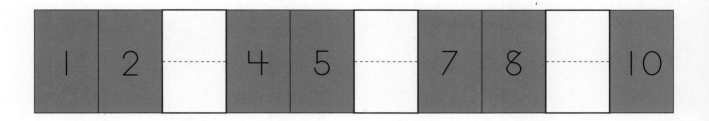

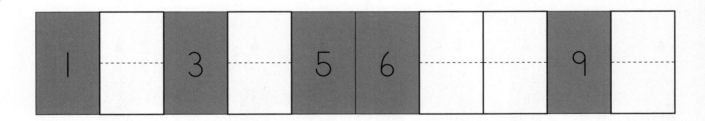

Circle the second from the right.

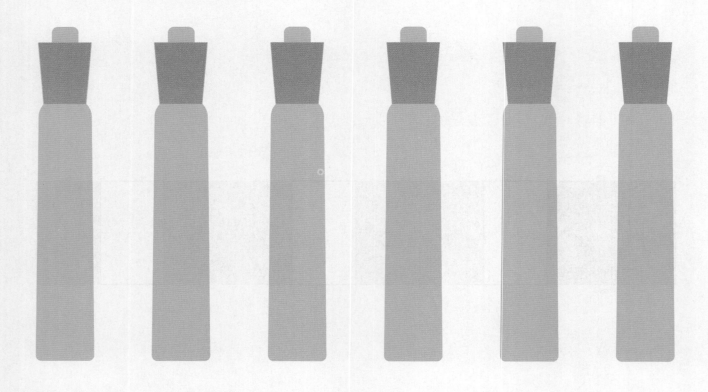

Circle the seventh from the left.

Circle the group that has 1 more in each box.

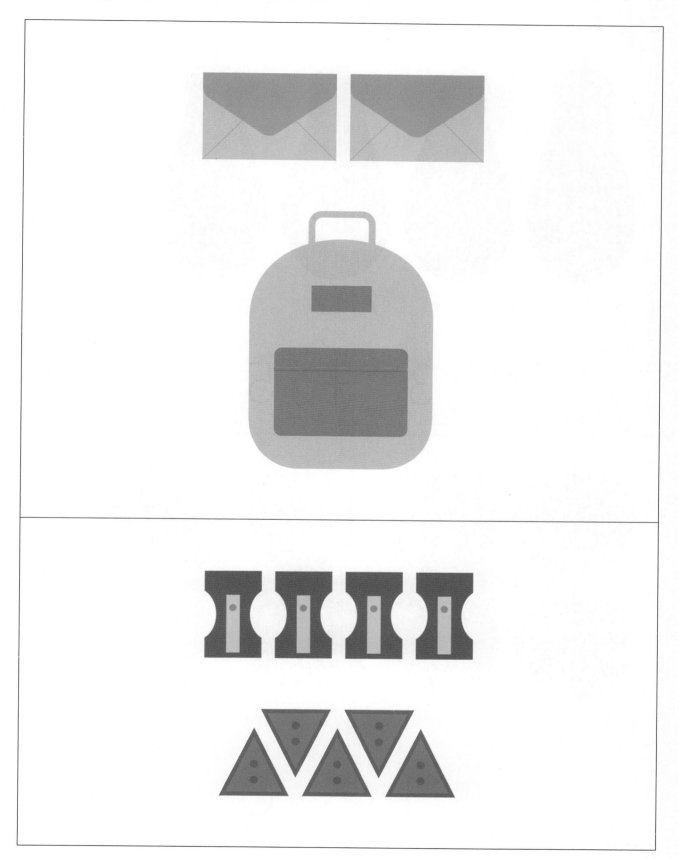

Draw one more and circle how many.

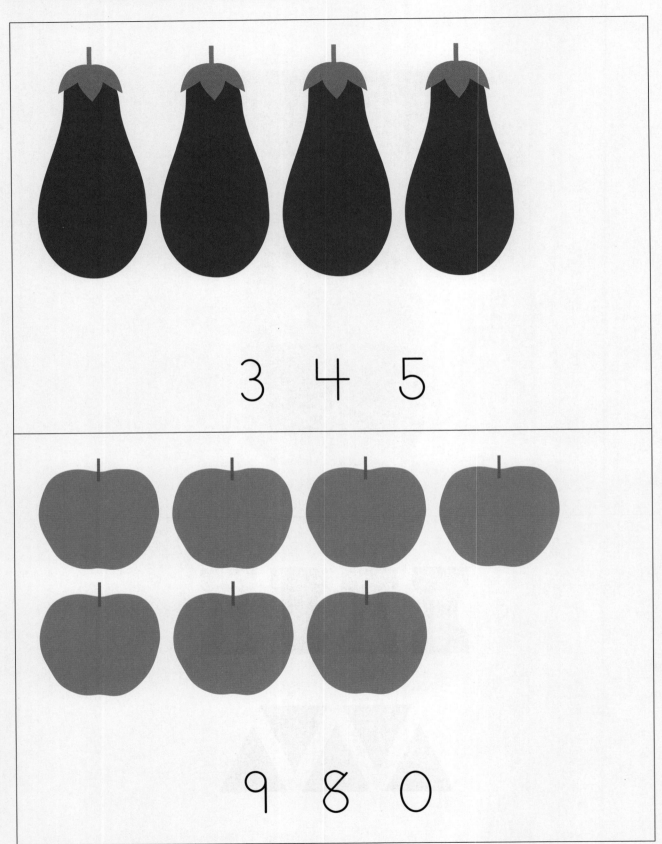

3 4 5

9 8 0

Chapter 3 Test 5

Test 6

Chapter 4 Shapes and Solids

(5 points each)

Circle the solid that is similar to the first object.

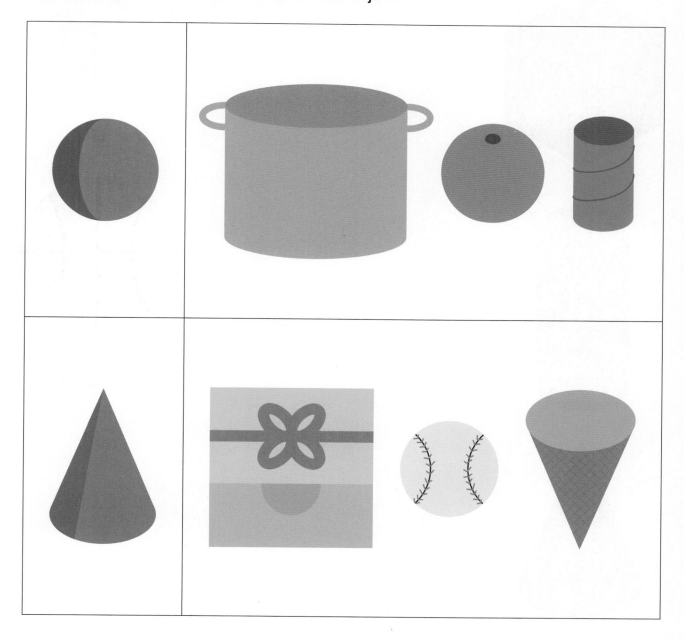

Match the solids.

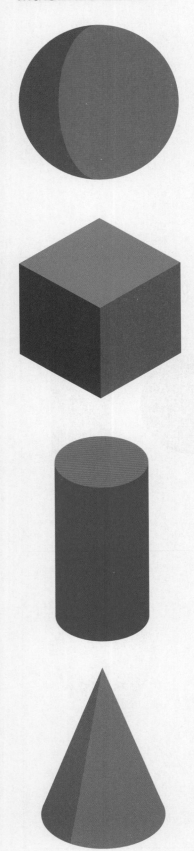

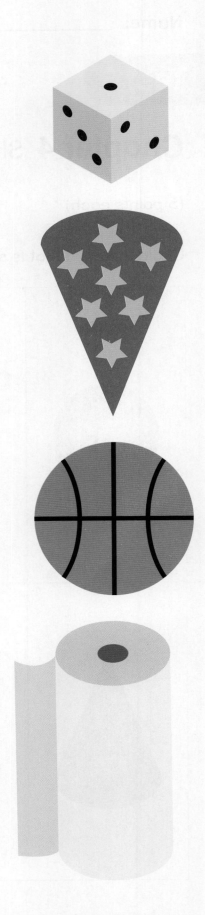

Color the closed shapes.

Circle the squares.

Match similar shapes.

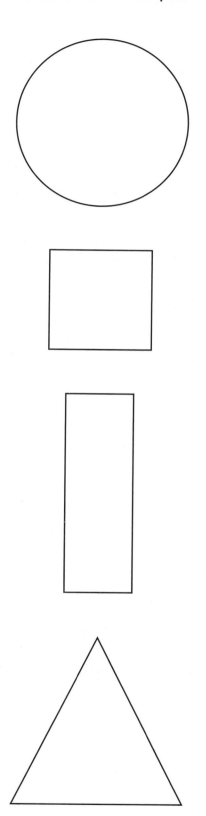
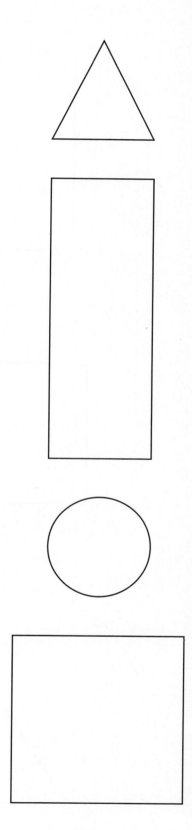

Color the triangles blue.

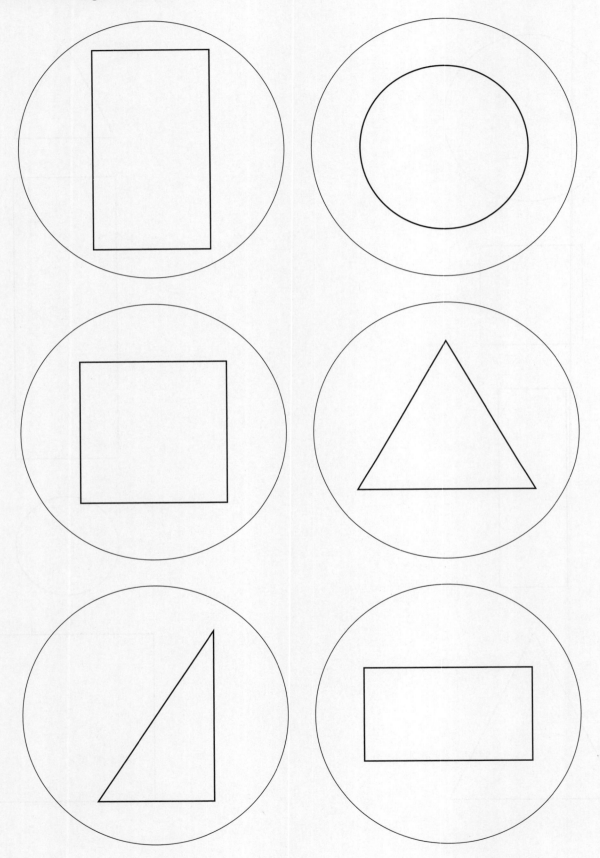

Test 7

Chapter 4 Shapes and Solids

(5 points each)

Cross out the shape that does not belong.

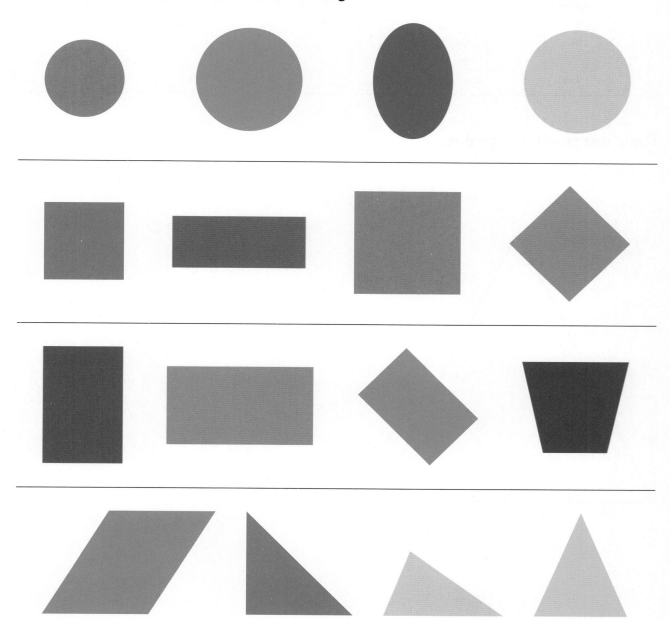

Draw a bigger circle.

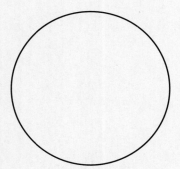

Draw two smaller triangles.

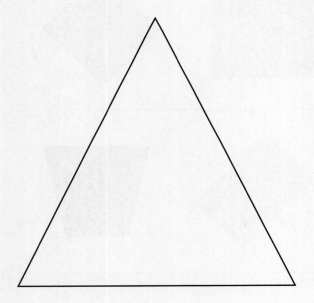

Color the biggest hexagon.
Circle the smallest square.

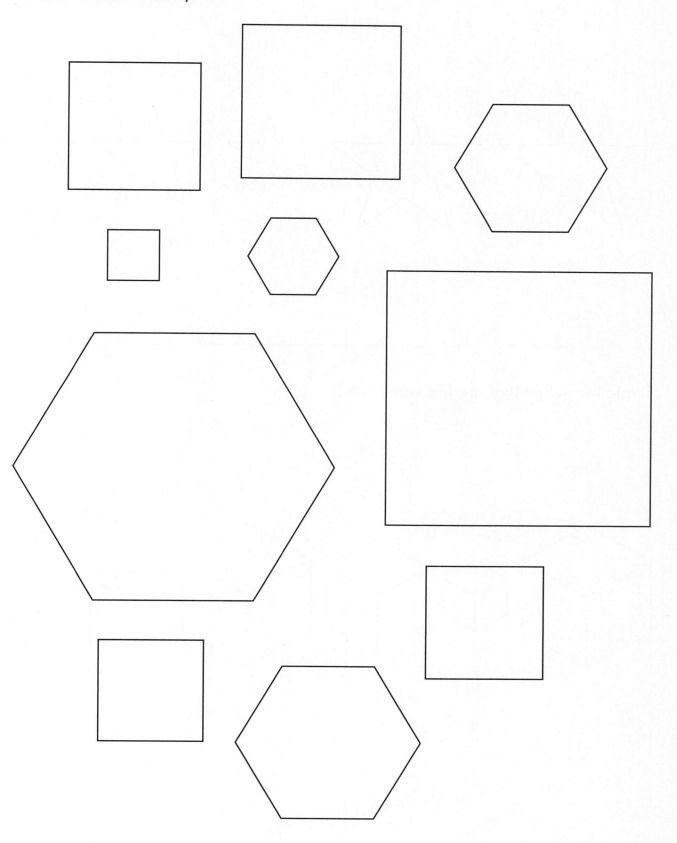

Circle the shapes that are the same size.

Circle the solids that are the same size.

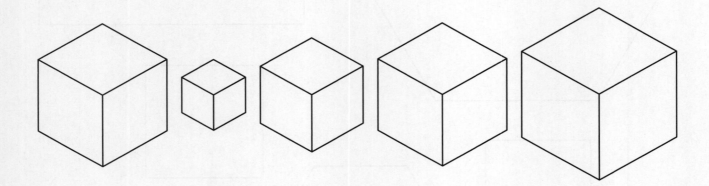

Draw the shape that completes the pattern.

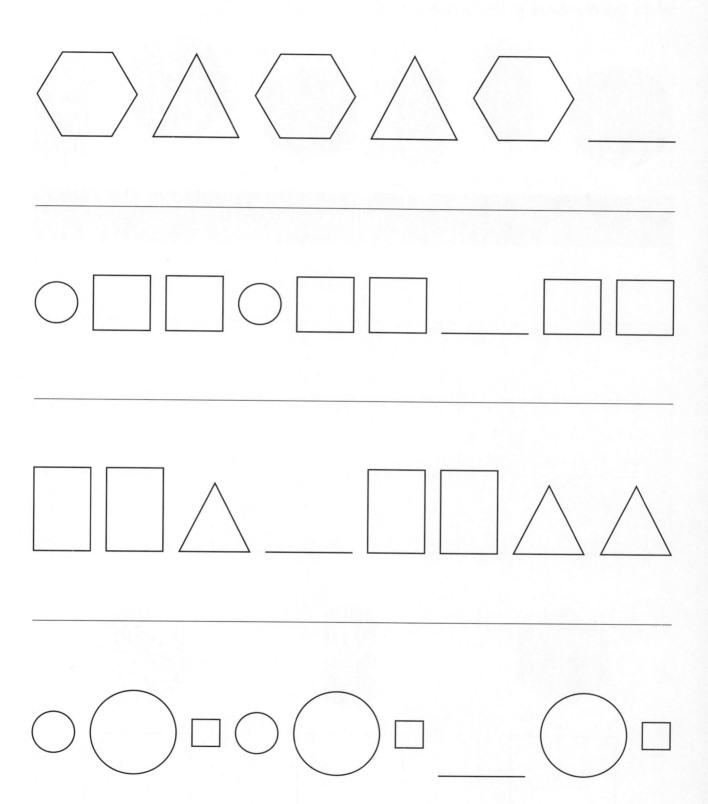

Count and color the graph.
Write the numbers in the boxes below.

Solids		

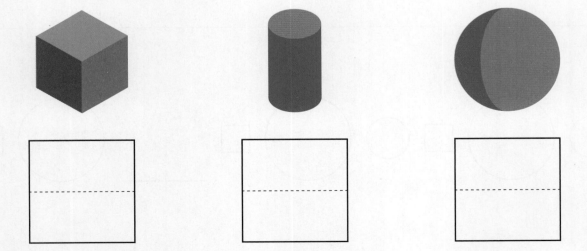

Test 8

Chapter 5 Compare Height, Length, Weight, and Capacity

(5 points each)

Circle the tallest tree.
Cross out the shortest tree.

Circle the shorter child in each box.

Chapter 5 Test 8

Circle the two things that are the same height in each row.

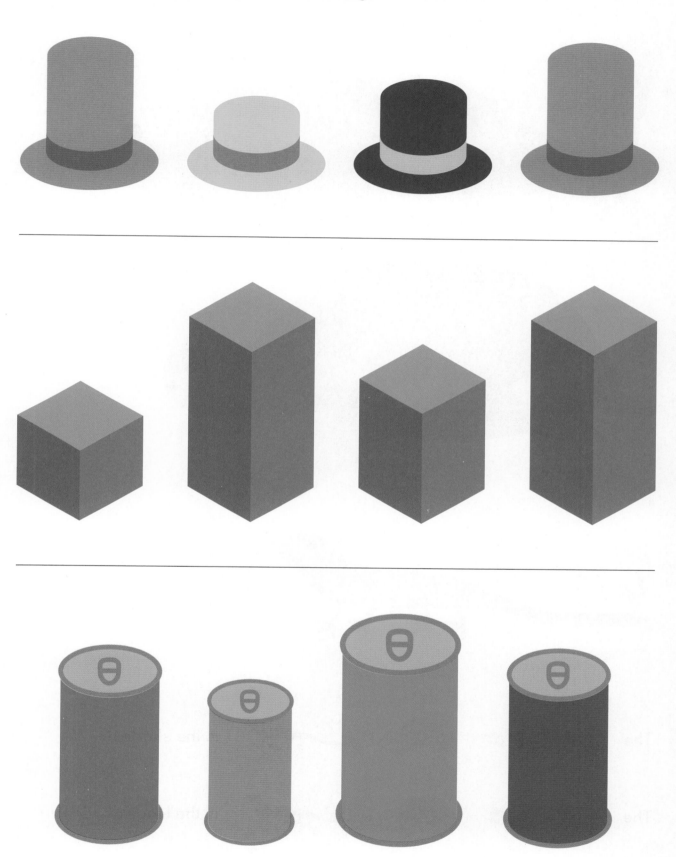

Circle the longest and the shortest animal below.

The    is the shortest.

The    is the longest.

Draw a shorter line.

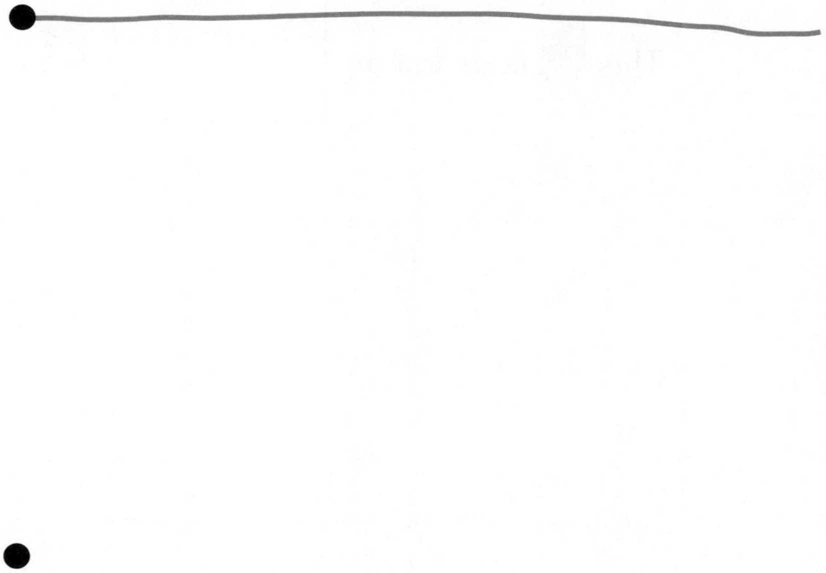

Count and write the number.

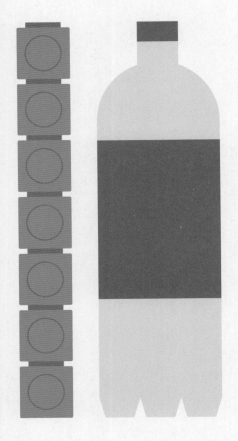

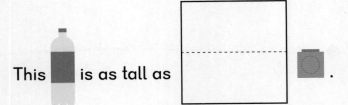

This is as tall as _____ .

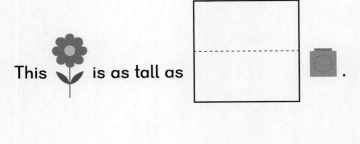

This is as tall as _____ .

Test 9

Chapter 5 Compare Height, Length, Weight, and Capacity

(5 points each)

Circle the heavier thing.

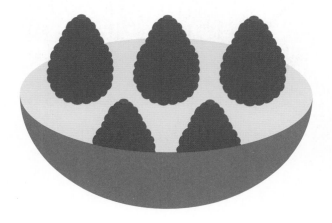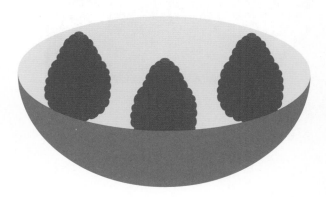

Circle the lighter one.

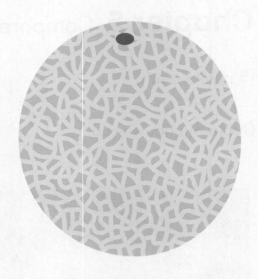

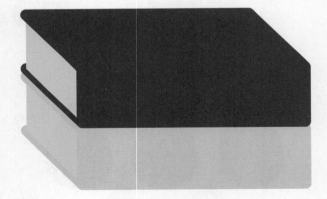

Circle the lighter thing.

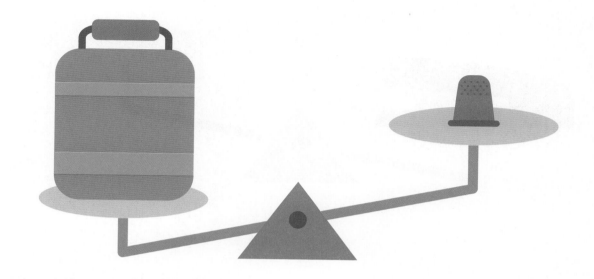

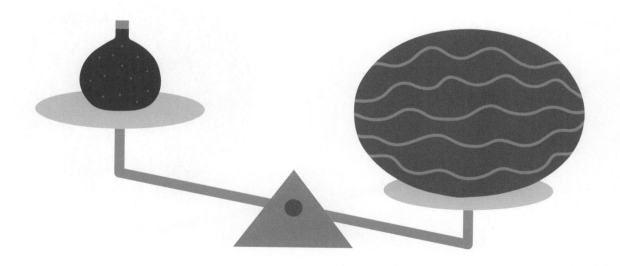

Circle the heavier thing.

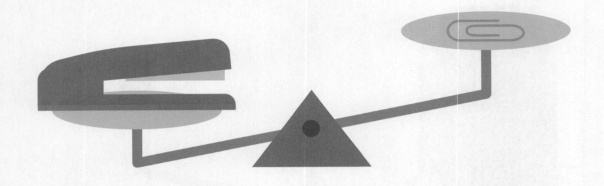

Circle the container that can hold more.

Circle the container that holds less.

Chapter 5 Test 9

Test 10

Chapter 6 Comparing Numbers Within 10

(5 points each)

Match each set of children with the same number of toys.

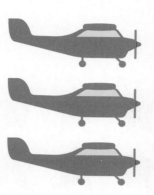

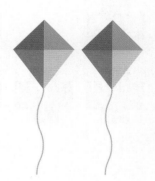

Match equal sets.

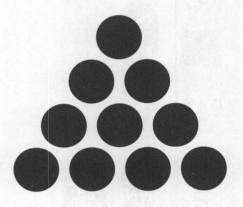

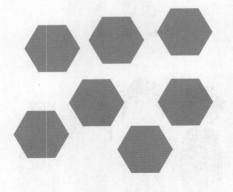

Circle the group that has more.

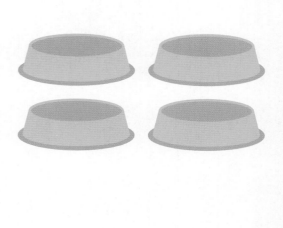

Circle the group that has fewer.

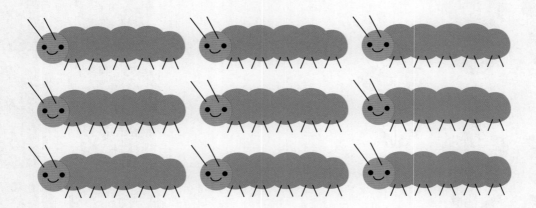

Chapter 6 Test 10

Cross out the bowl that has fewer strawberries.

Count and write the numbers.

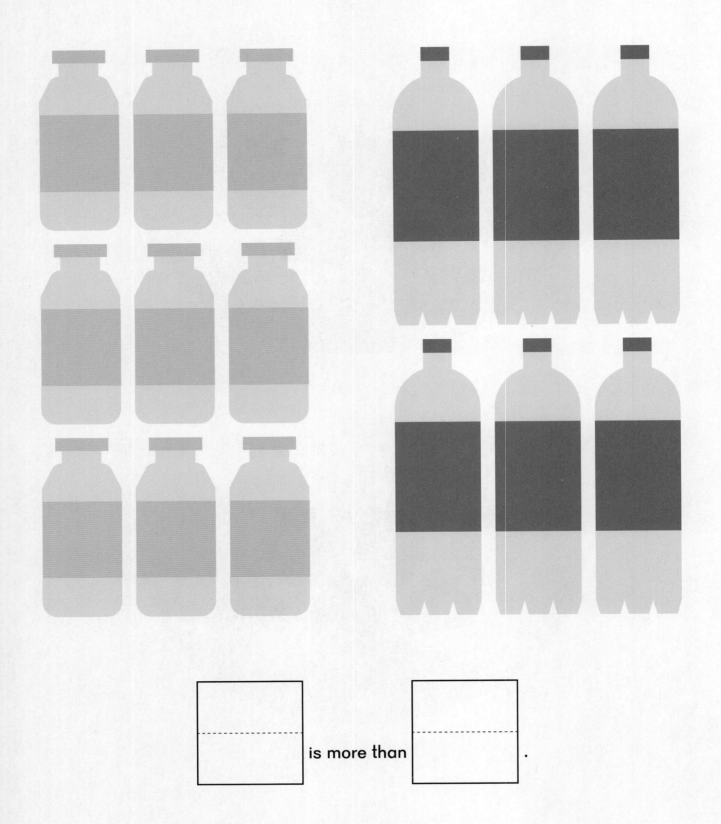

is more than .

Count and write the numbers.

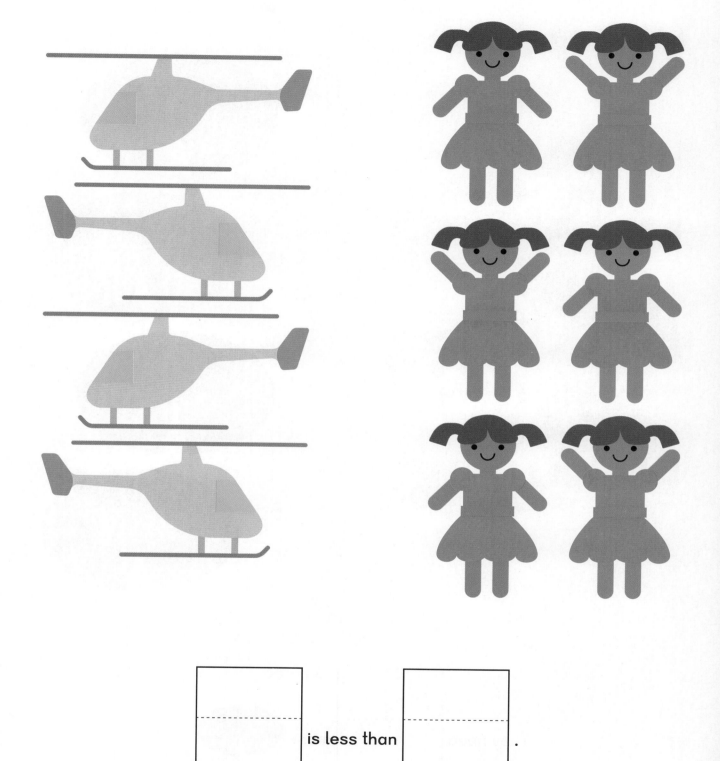

	is less than	

Write the number in the box.

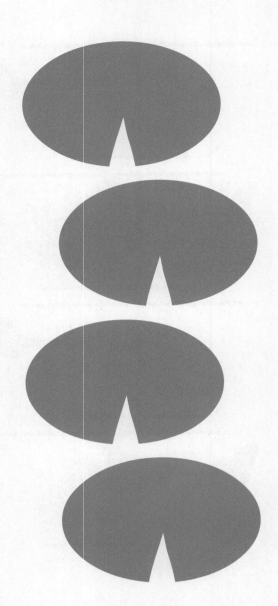

They need [ ] more .

Test 11

Chapter 7 Numbers to 20

(5 points each)

Circle the groups of 10.

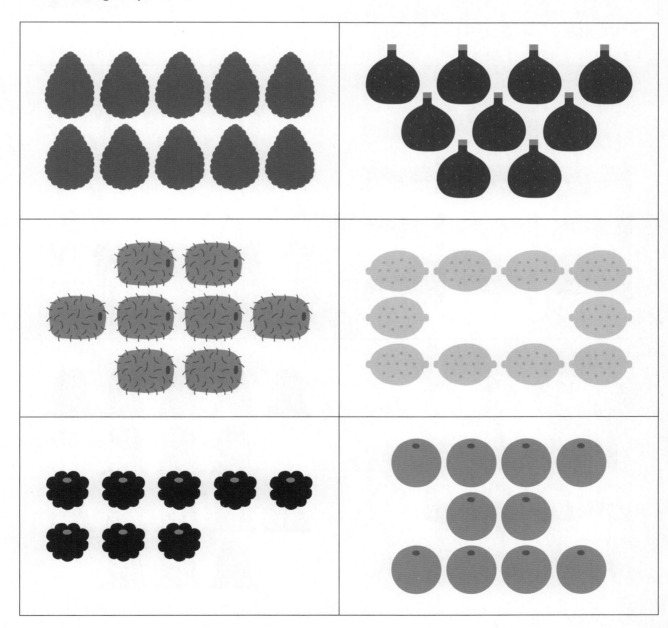

Count and circle 10.

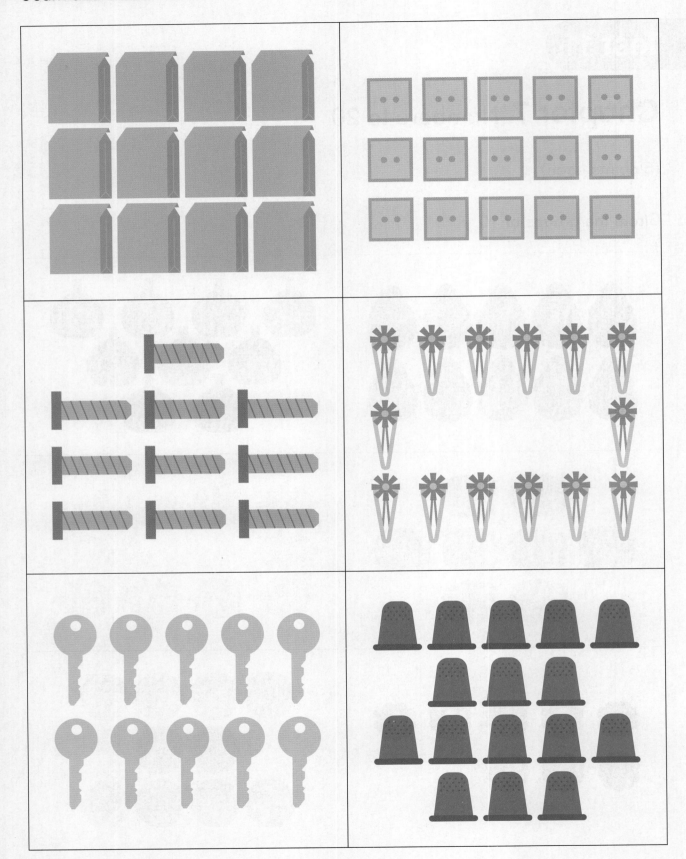

Count and circle 10.
Write the numbers.

Tens	Ones

Tens	Ones

Tens	Ones

Count the dots and write the numbers.

Match.

6

1

10

5

five

ten

six

one

Count the dots.
Circle the correct word.

(domino: 2 dots and 3 dots)	four five
(domino: 4 dots and 4 dots)	nine eight
(domino: blank)	zero one
(domino: 3 dots and 3 dots)	six seven

Test 12

Chapter 7 Numbers to 20

(5 points each)

Match.

thirteen   ten and four

twelve   ten and five

fifteen   ten and three

fourteen   ten and two

Write the numbers.

eleven		Tens	Ones	
sixteen		Tens	Ones	
twelve		Tens	Ones	
nineteen		Tens	Ones	

Circle the correct word.

●●●●● ●●●●● ●●●● ●●	eighteen thirteen
●●●●● ●●●●● ●●●●● ●●●●●	twelve twenty
●●●●● ●●●●● ●●●● ●●	seventeen seven
●●●●● ●●●●● ●●●●●	eleven fifteen

Fill in the missing numbers.

| 13 | | 15 | | 17 | |

| 20 | | 18 | 17 | 16 | |

Draw one more apple.

Write the number that is one more.

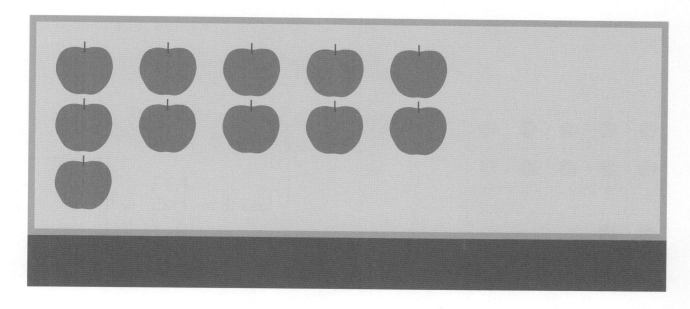

| more than | | is [ ] .

| more than | 7 is [ ] .

Cross one off.
Write the number that is one less.

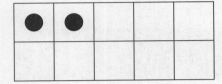

I less than 12 is [ ] .

I less than 20 is [ ] .

Chapter 8 Number Bonds

(5 points each)

How many in all?
Write the numbers.

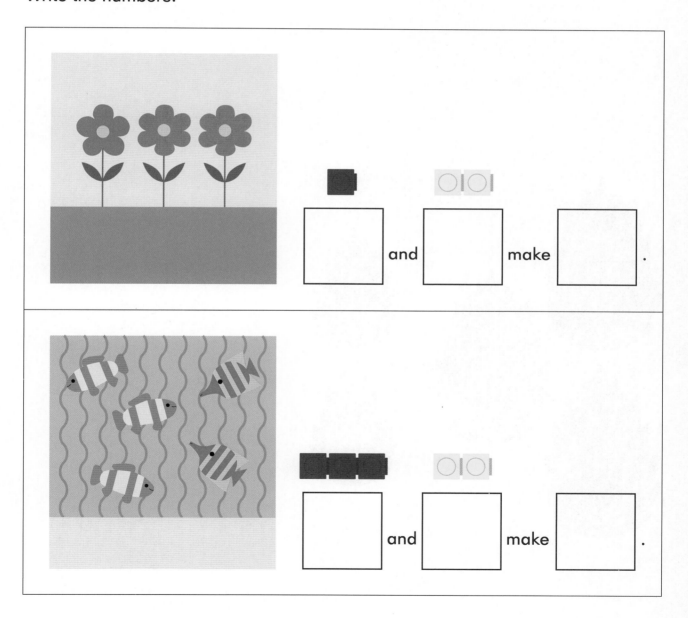

Complete the number bonds.

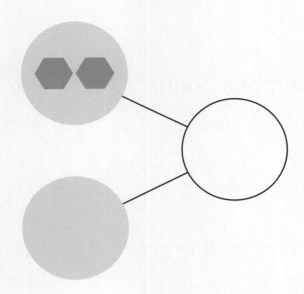

Complete the number bonds.

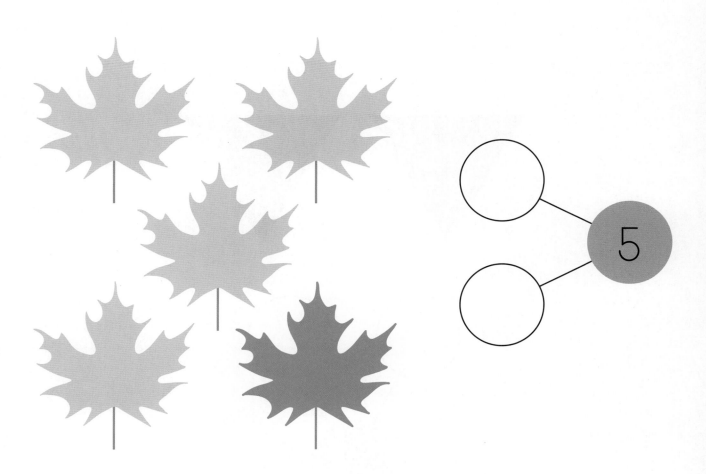

Complete the number bonds.

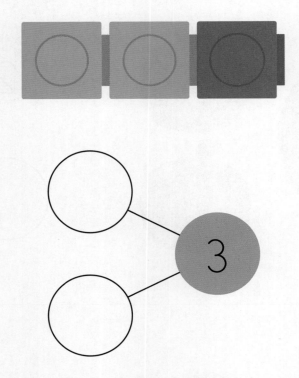

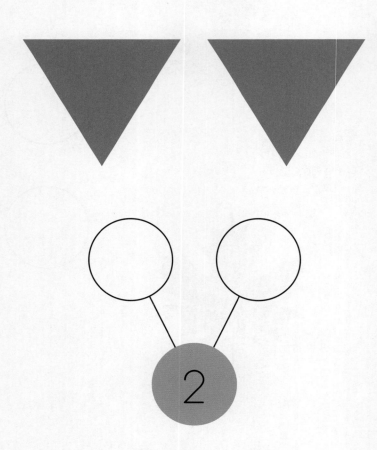

Complete the number bonds.

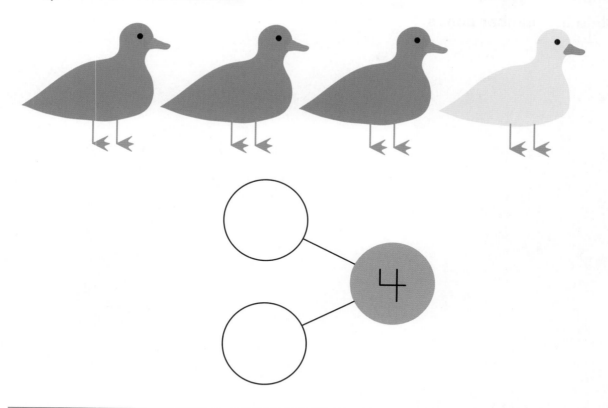

Draw more to make 5.
Complete the number bonds.

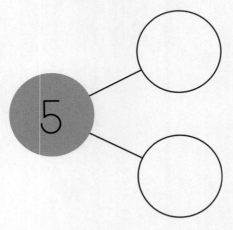

Test 14

Chapter 8 Number Bonds

(5 points each)

Complete the number bonds.

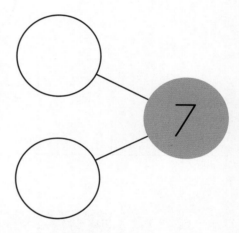

Complete the number bonds.

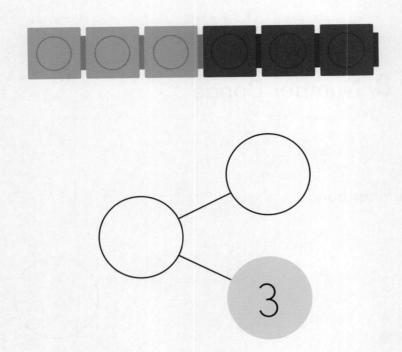

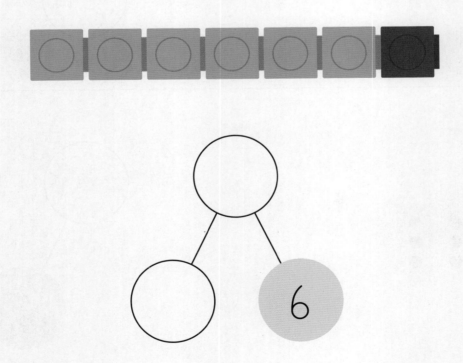

Complete the number bonds.

Draw more to make 8.
Complete the number bonds.

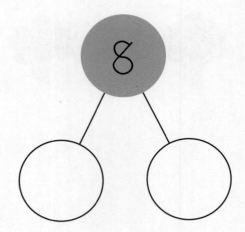

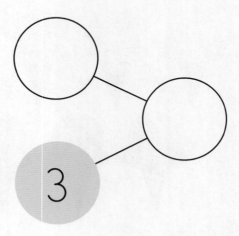

Complete the number bonds.

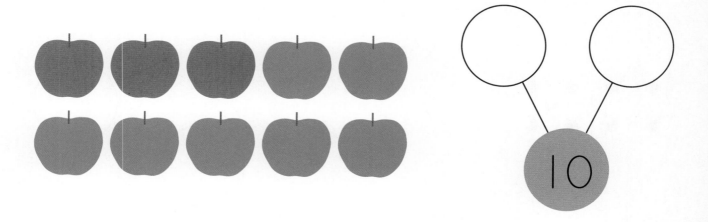

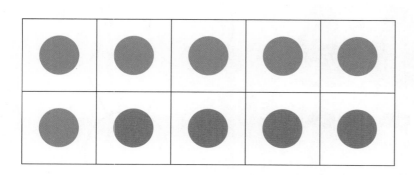

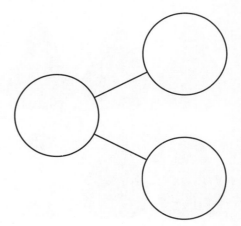

Complete the number bonds.

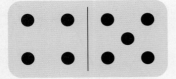

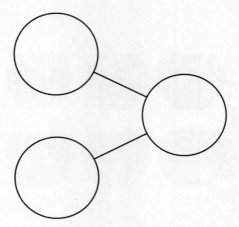

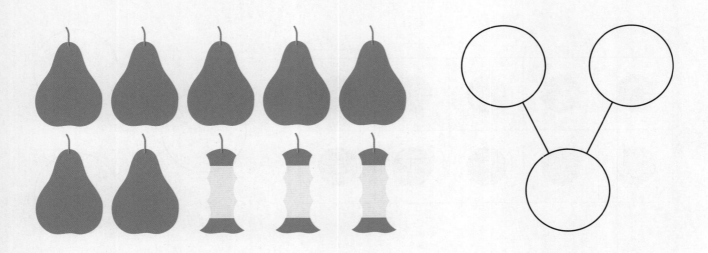

Test 15

Chapter 9 Addition

(5 points each)

Complete the number bonds.
Complete the number sentences.

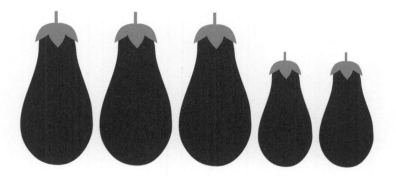

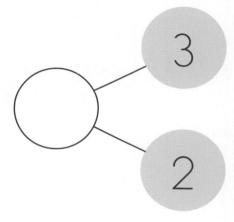

$$\boxed{} = 3 + 2$$

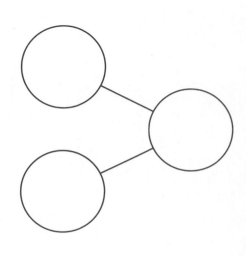

$$\boxed{} + \boxed{} = \boxed{}$$

Use a different color to color 2 more ribbons.
Complete the number sentence.

4 + 2 =

Color the correct number of balloons.
Complete the number sentences.

5 + 0 = ☐

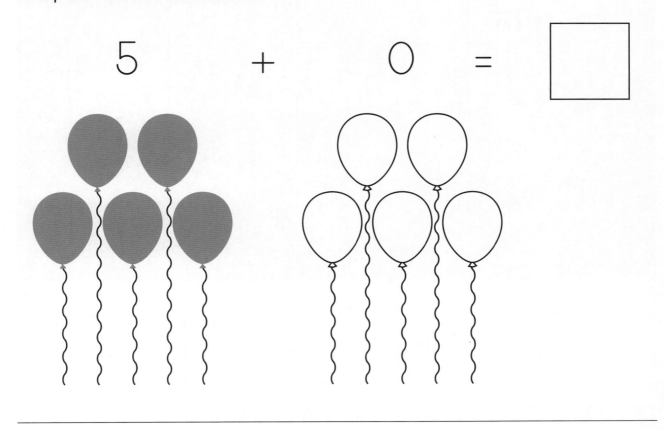

4 + 3 = ☐

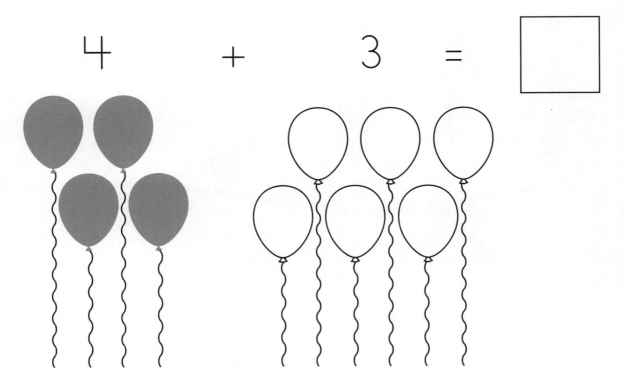

Complete the number sentences.

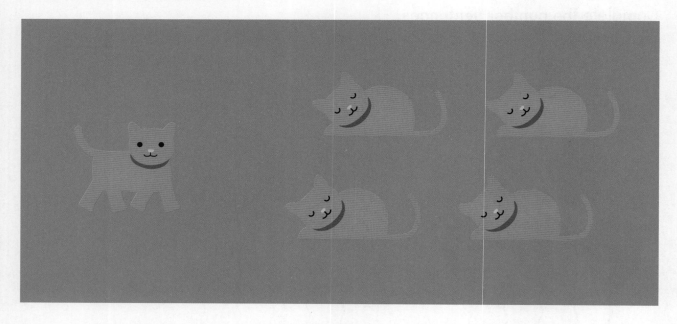

$1 + \boxed{} = \boxed{}$

$\boxed{} + 2 = \boxed{}$

Count on and complete the number sentences.

$$6 \quad + \quad \boxed{} \quad = \quad \boxed{}$$

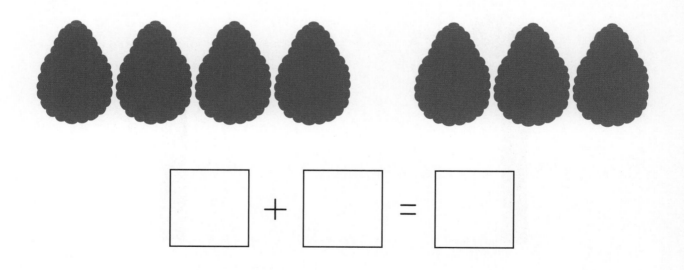

$$\boxed{} \quad + \quad \boxed{} \quad = \quad \boxed{}$$

Draws apples on the tree on the right.
Count on and write the number.

6 + 3 = ☐

Test 16

Chapter 9 Addition

(5 points each)

Complete the number sentences.

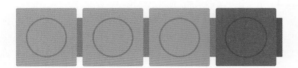

$$3 + \boxed{} = 4$$

$$4 = \boxed{} + 0$$

Complete the number sentences.

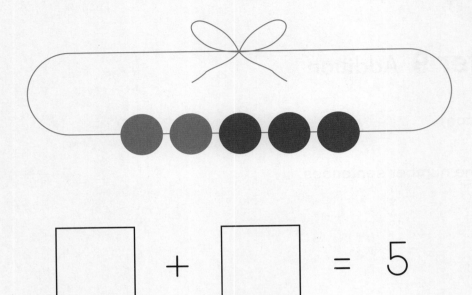

$$\boxed{} + \boxed{} = 5$$

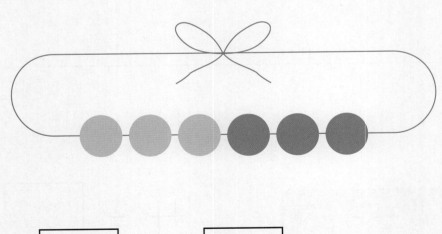

$$\boxed{} = \boxed{} + 3$$

Complete the number sentences.

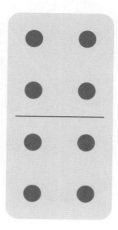

$3 + \boxed{} = \boxed{}$

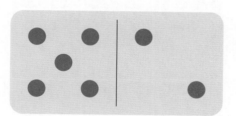

$\boxed{} + 2 = \boxed{}$

Complete the number sentences.

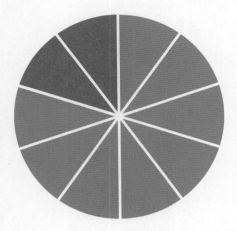

$10 = \boxed{} + \boxed{}$

$\boxed{} + \boxed{} = 9$

Complete the number sentences.

$$\boxed{} + \boxed{} = \boxed{}$$

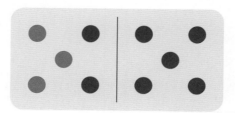

$$\boxed{} + \boxed{} = \boxed{}$$

Complete the number sentences.

$$7 + \boxed{} = 9$$

$$9 = \boxed{} + 2$$

$$4 + 6 = \boxed{}$$

$$10 = \boxed{} + 6$$

Test 17

Chapter 10 Subtraction

(5 points each)

Cross off what Mei eats.
Write the number of ice cream cones that remain.

Mei eats 1 .

remain.

Cross off the crabs that swim away.
Write the number of crabs that remain.

3  swim away.

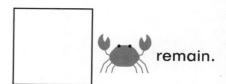
remain.

Complete the number sentences.

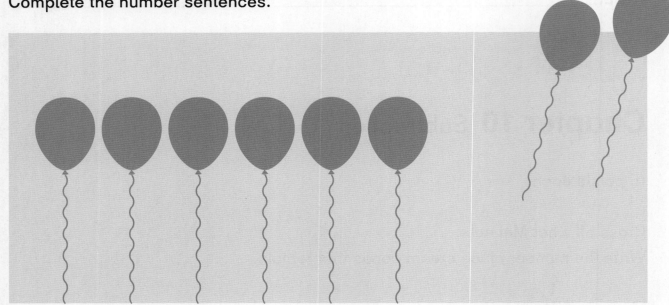

$$8 - 2 = \boxed{}$$

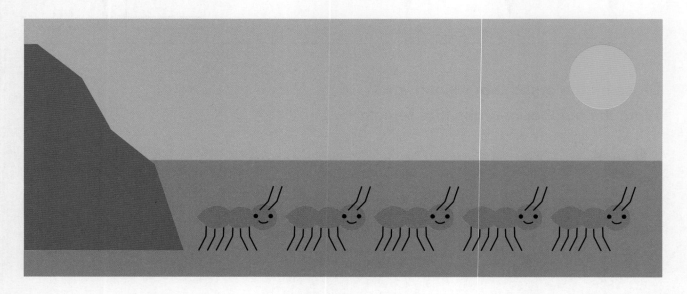

$$5 - 5 = \boxed{}$$

Complete the number sentences.

$7 - \boxed{} = \boxed{}$

$\boxed{} - 1 = \boxed{}$

Complete the number sentences.

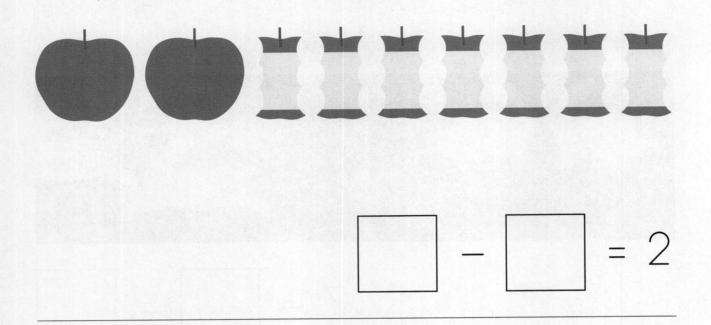

$$\boxed{} - \boxed{} = 2$$

$$\boxed{} - \boxed{} = 0$$

Complete the number sentences.

Complete the number bonds and number sentences.

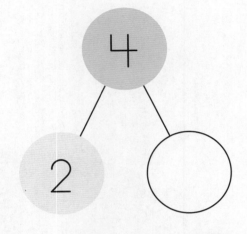

$$4 - 2 = \boxed{}$$

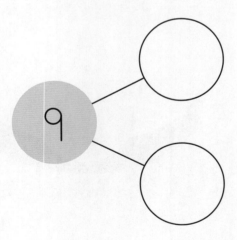

$$9 - 3 = \boxed{}$$

Test 18

Chapter 10 Subtraction

(5 points each)

Complete the number sentences.

$4 - 2 =$ ☐

$9 - 3 =$ ☐

Complete the number sentences.

$5 - 2 = \boxed{}$

$5 - \boxed{} = 1$

Complete the number sentences.

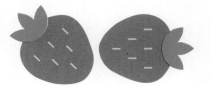

$8 \; - \; \boxed{} \; = \; \boxed{}$

$8 \; = \; \boxed{} \; - \; \boxed{}$

How many remain?
Write the numbers.

$$5 - 1 = \boxed{}$$

$$\boxed{} = 7 - 4$$

Cross off the correct number of things.

Complete the number sentences.

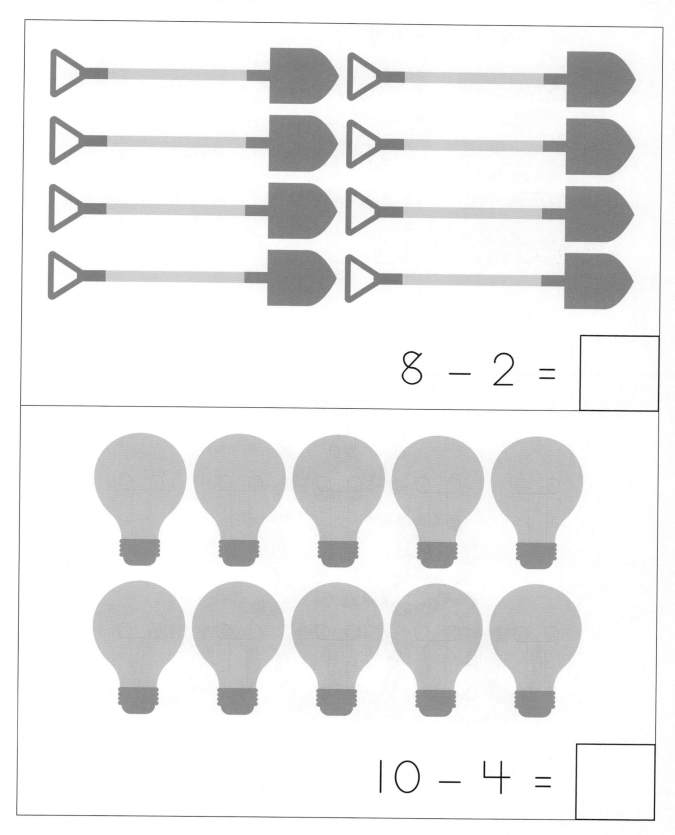

8 − 2 =

10 − 4 =

Complete the number sentences.

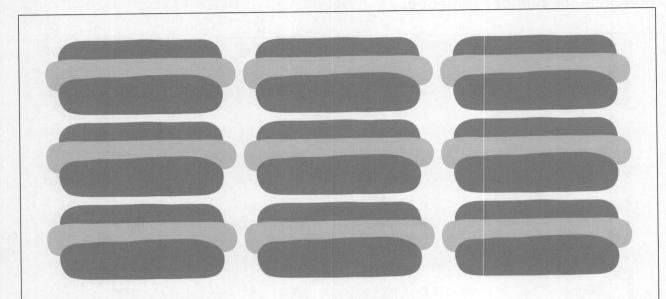

$$9 - 5 = \boxed{}$$

$$6 - 6 = \boxed{}$$

Test 19

Chapter 11 Addition and Subtraction

(5 points each)

Add or subtract.
Complete the number sentences.

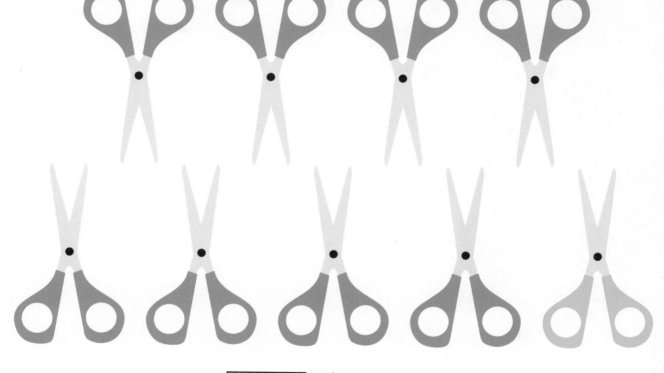

$8 + 1 = \boxed{}$ $\quad$ $1 + 8 = \boxed{}$

$9 - 1 = \boxed{}$ $\quad$ $9 - 8 = \boxed{}$

Add or subtract.
Complete the number sentences.

$5 + 2 =$ ☐ | $2 + 5 =$ ☐

$7 - 2 =$ ☐ | $7 - 5 =$ ☐

Complete the number sentences.

Alex and Emma have ☐ 🎈 altogether.

Complete the number sentences.

There are 7 foxes.

2 foxes are joining them.

How many  are there altogether?

□ ○ □ = □

There are □ 🦊 altogether.

Complete the number sentences.

There were 6 mice eating in all.
3 ran away.
How many  remain?

□ ◯ □ = □

□ 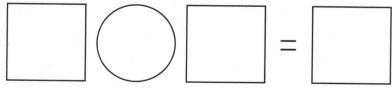 remain.

Complete the number sentences.

How many peppers are on the plate?

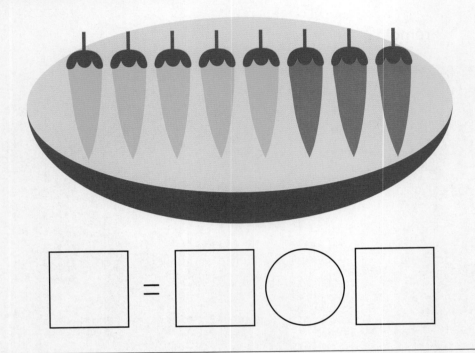

$$\boxed{} = \boxed{} \bigcirc \boxed{}$$

How many bananas are green?

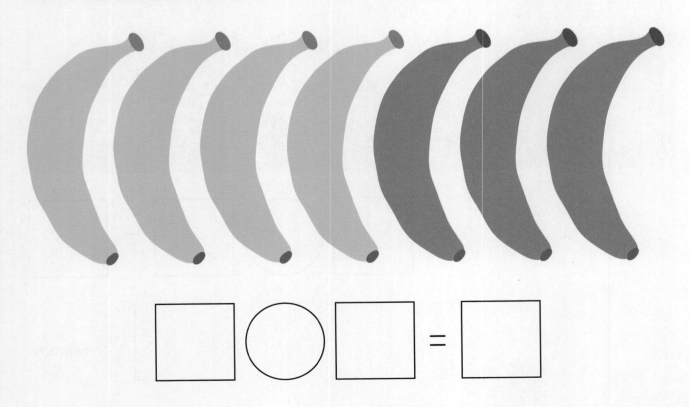

$$\boxed{} \bigcirc \boxed{} = \boxed{}$$

Test 20

Chapter 12 Numbers to 100

(5 points each)

Match.

eighty

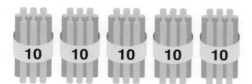

twenty

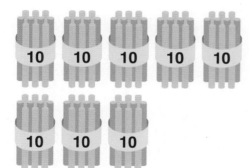

60
sixty

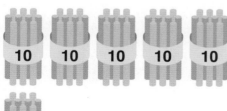

50
fifty

Count by 10.
Circle the correct number of cubes.

30
3 tens

100
10 tens

70
7 tens

10
1 ten

Count by tens.

Write the number of tens.

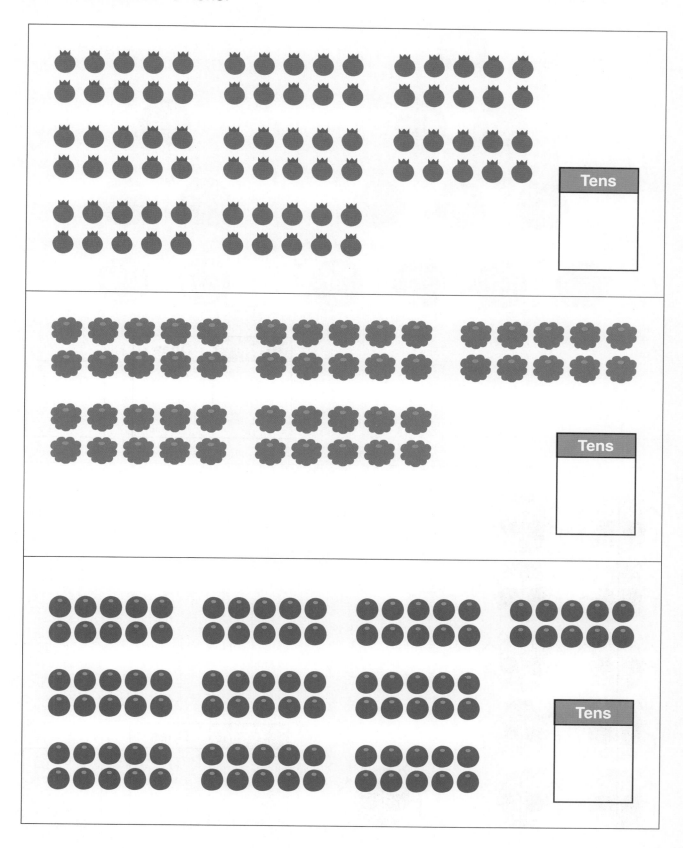

Tens

Tens

Tens

Count.
Write the numbers.

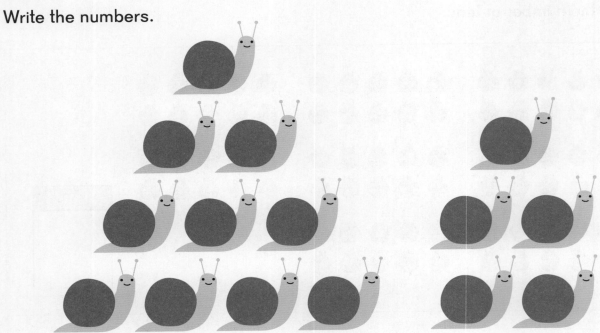

Tens	Ones

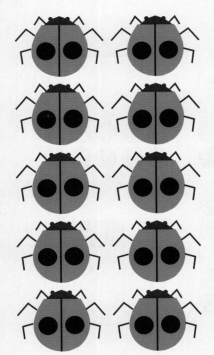

Tens	Ones

Match.

26

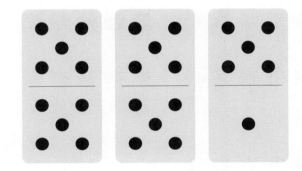

34

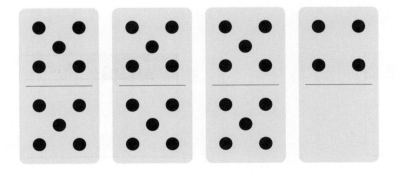

40

Count.

Write the numbers.

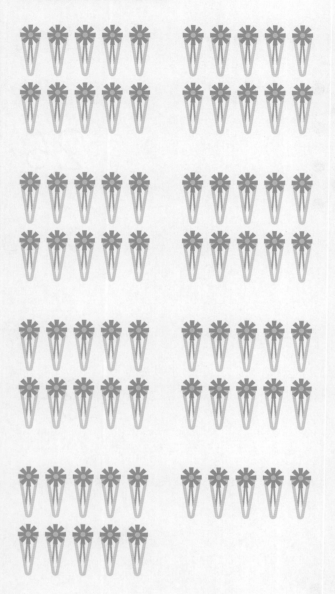

Tens	Ones

Tens	Ones

Test 21

Chapter 12 Numbers to 100

(5 points each)

Circle groups of 10.
Count.
Write the numbers.

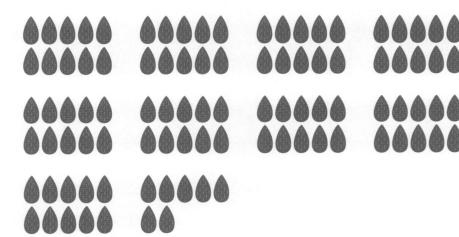

Tens	Ones

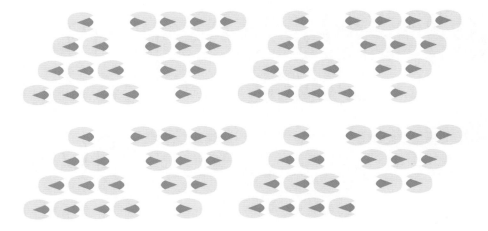

Tens	Ones

Count.

Circle the correct numbers.

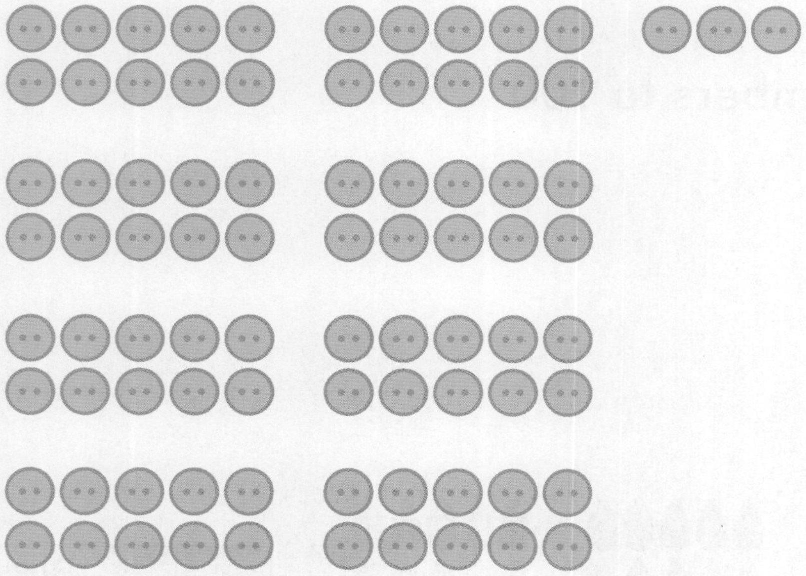

85

80

83

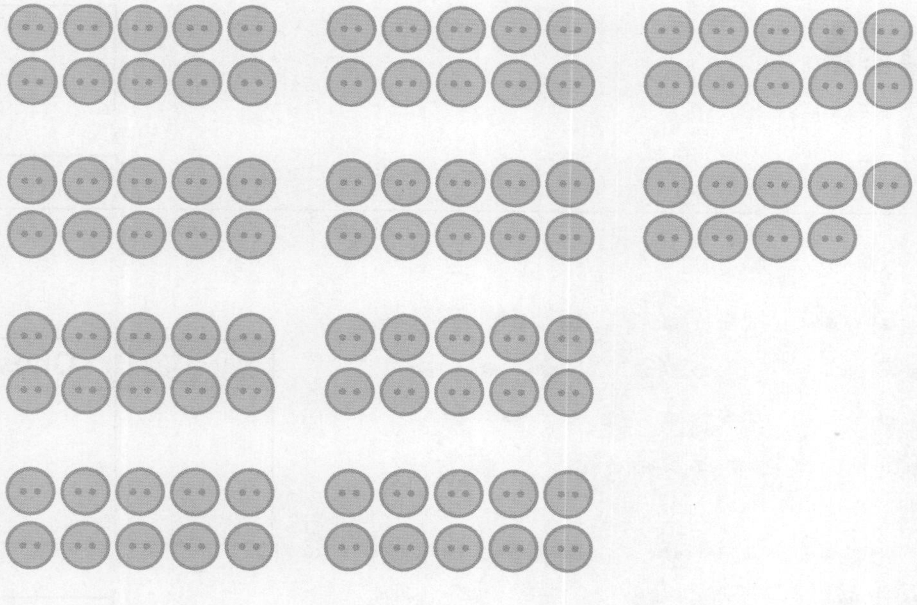

89

99

90

Count.
Write the numbers.

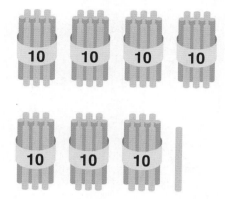

Tens	Ones

☐

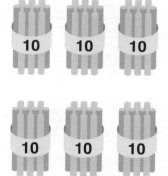

Tens	Ones

☐

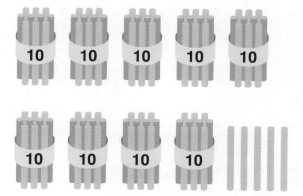

Tens	Ones

☐

Fill in the missing numbers.

1		3	4	5	6	7	8	9	
11	12	13		15	16	17	18	19	20
21	22	23	24	25			28	29	30
	32	33	34	35	36	37	38	39	40
41	42		44	45	46	47	48	49	
51	52	53	54	55	56	57	58		60
	62	63	64	65	66	67	68	69	70
71	72	73	74	75	76	77	78	79	80
81	82	83	84	85	86	87		89	90
91	92	93	94		96	97	98	99	

Count by fives.
Write the numbers.

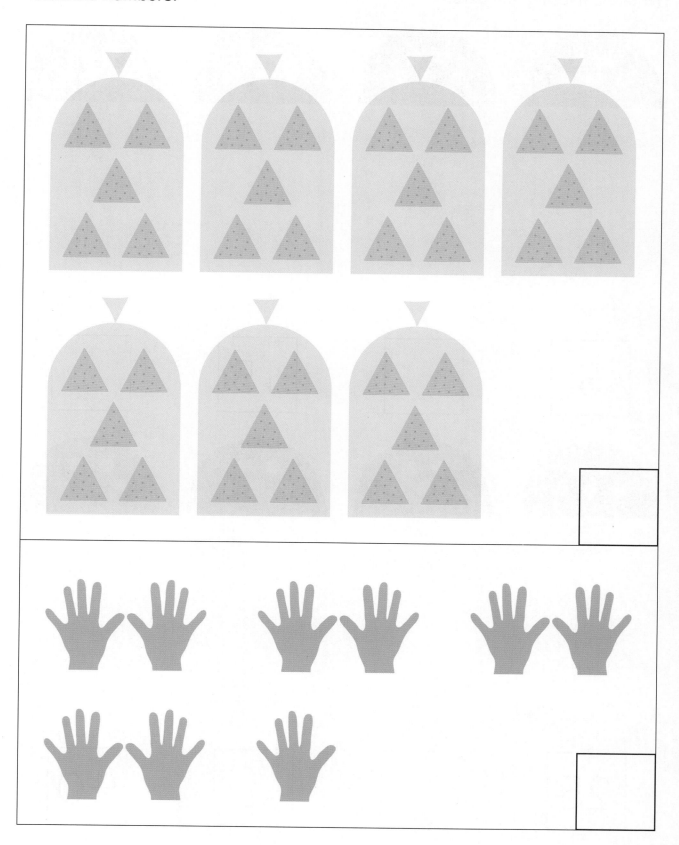

Count by fives.
Write the missing numbers.

5

15

25

40

Test 22

Chapter 13 Time

(5 points each)

Write the missing numbers.

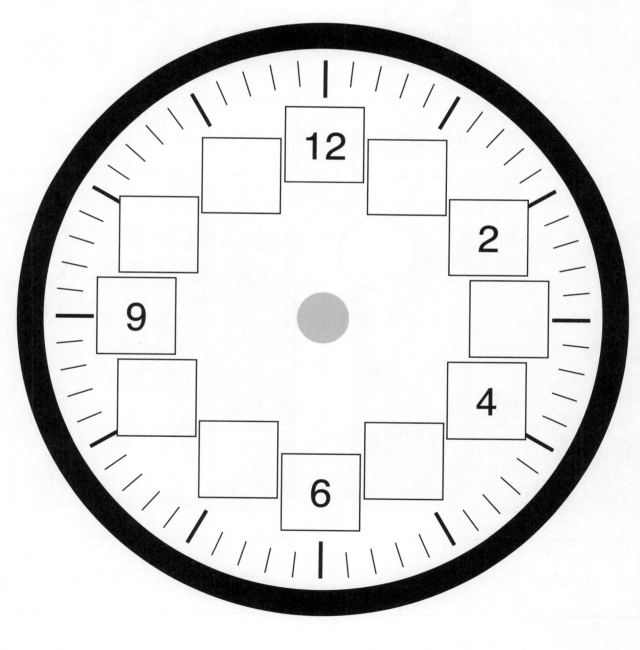

Circle the correct time.

| 7 o'clock | 8 o'clock |

| 12 o'clock | 1 o'clock |

| 8 o'clock | 4 o'clock |

| 3 o'clock | 9 o'clock |

Match.

2 o'clock

11 o'clock

6 o'clock

10 o'clock

Write the time.

o'clock

o'clock

:00

:00

Draw the hour hand.

3 o'clock

6 o'clock

Draw the clock hands.

1:00

7:00

Test 23

Chapter 14 Money

(5 points each)

Match.

Penny

Dime

Nickel

Quarter

Match.

1¢

5¢

10¢

25¢

Count and write.

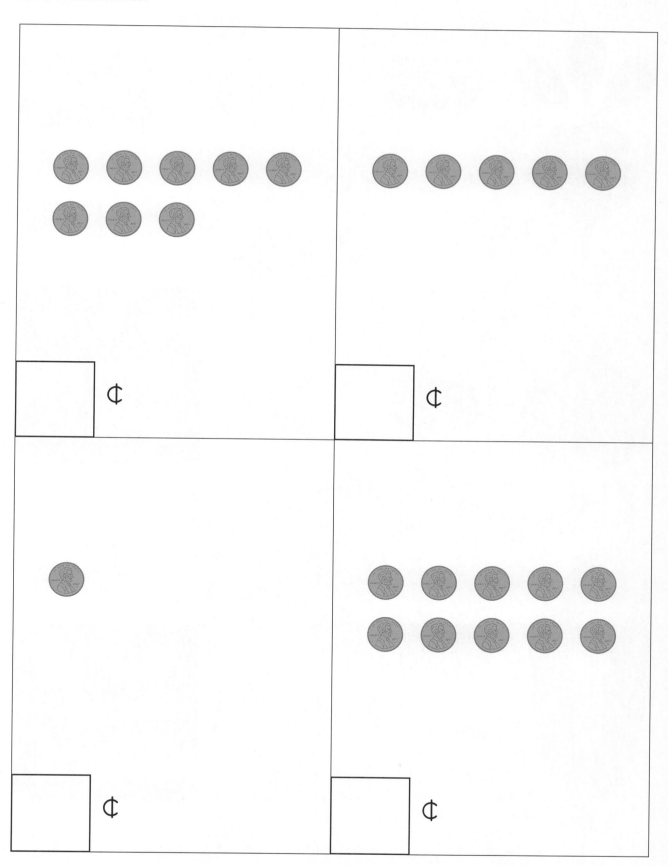

¢

¢

¢

¢

Match.

70¢

20¢

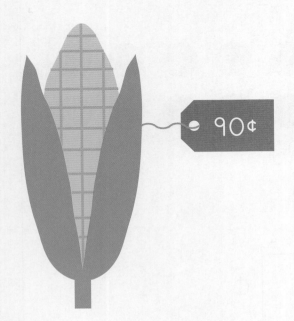

90¢

Count and write.

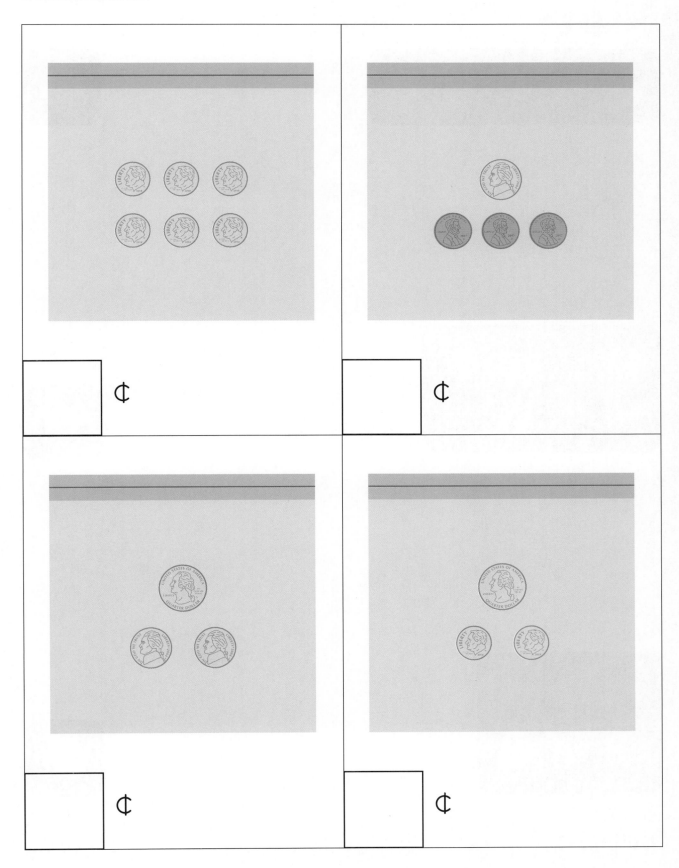

¢

¢

¢

¢

Match.

Answer Key KA

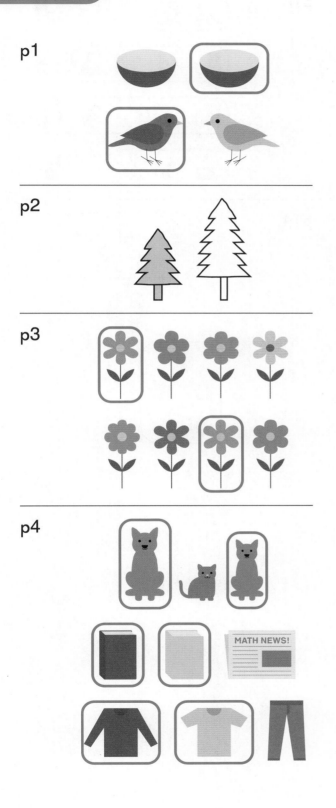

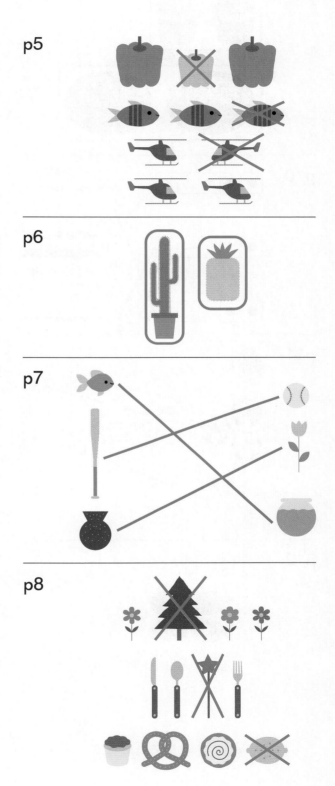

p9

p10

p11

p12

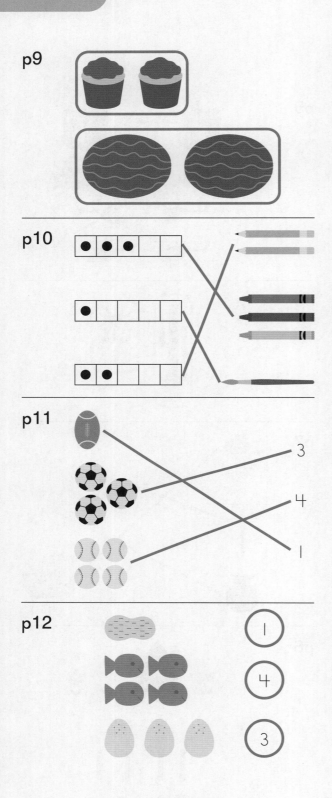

p13

2

4

5

3

p14

5

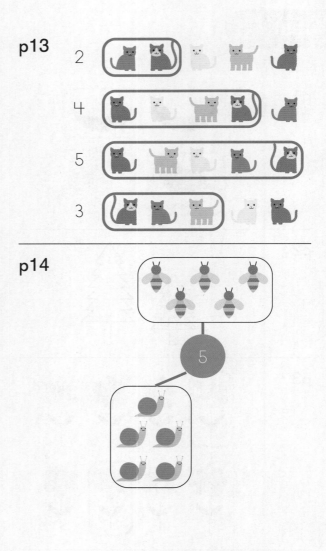

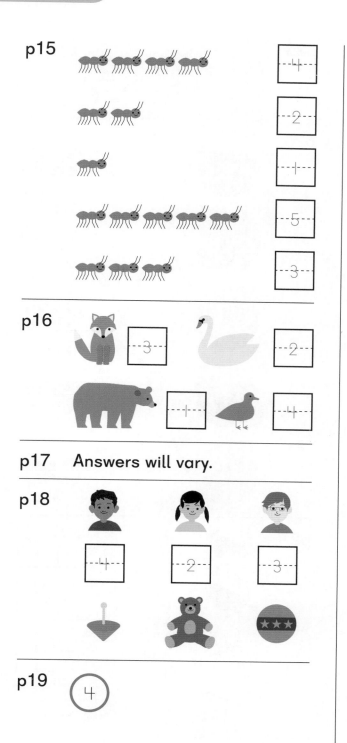

p15

4

2

1

5

3

p16

3

2

1

4

p17 Answers will vary.

p18

4

2

3

p19 4

p20

2

4

3

5

1

Test 4

p21

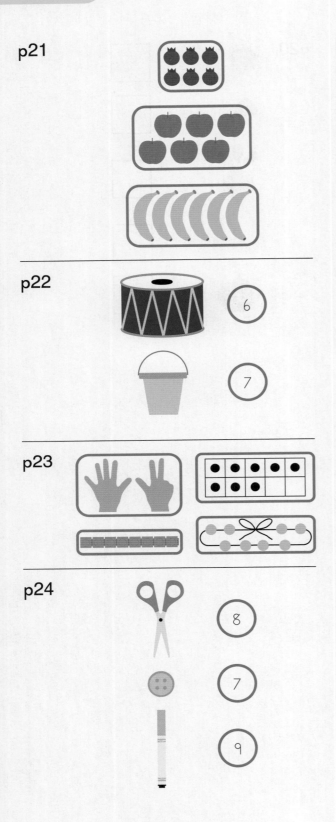

p22

p23

p24

p25

p26

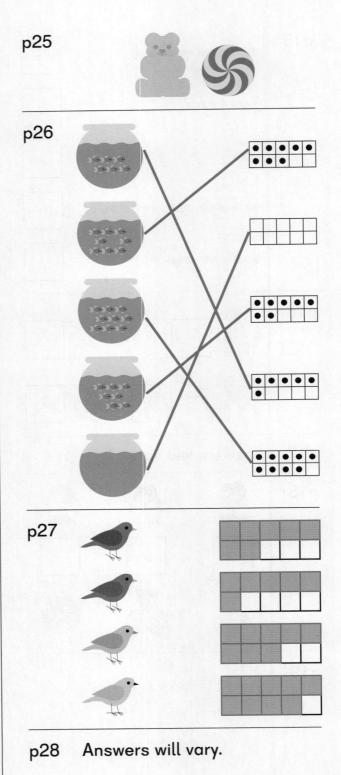

p27

p28   Answers will vary.

Test 5

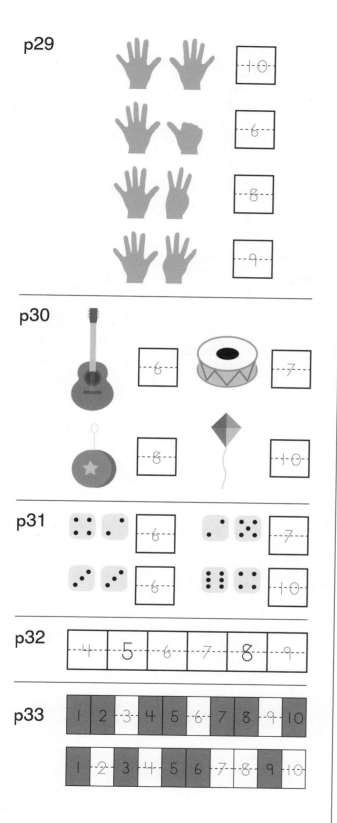

p29

p30

p31

p32

| | 5 | 6 | 7 | 8 | 9 |

p33

| 1 | 2 | 3 | 4 | 5 | 6 | 7 | 8 | 9 | 10 |

| 1 | 2 | 3 | 4 | 5 | 6 | 7 | 8 | 9 | 10 |

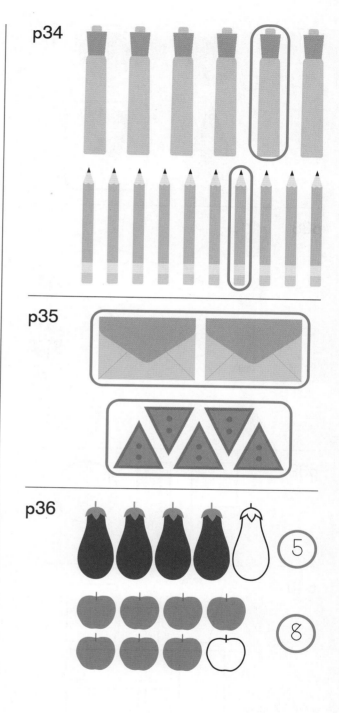

p34

p35

p36

⑤

⑧

p37

p38

p39

p40

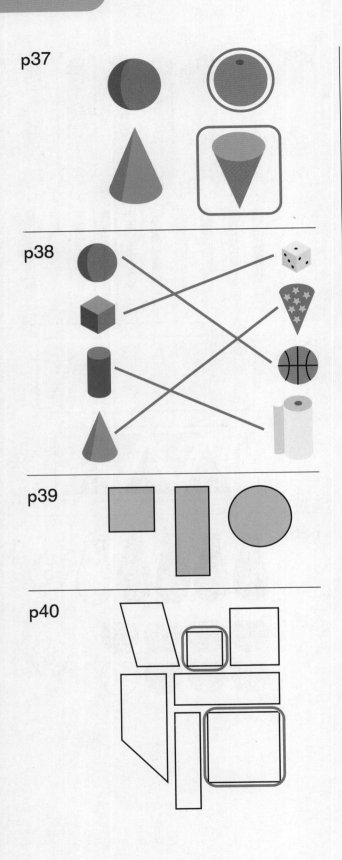

p41

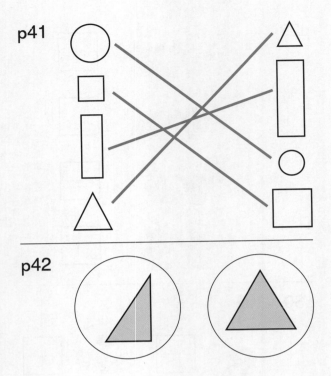

p42

p43

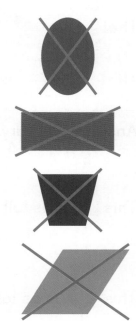

p47

p44   Answers will vary.

p45

p46

p48

Solids		

1   3   2

Test 8

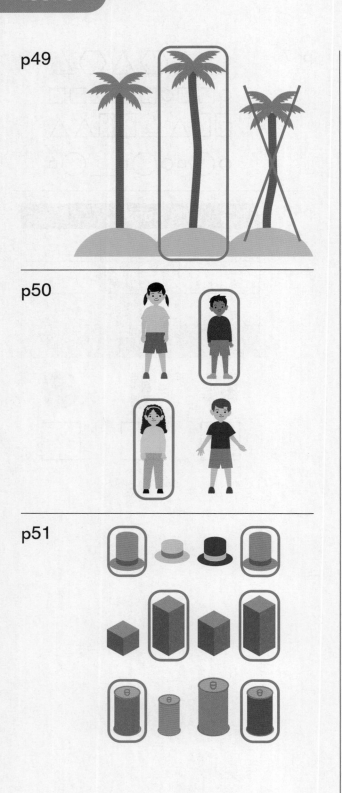

p49

p50

p51

p52 The 🦎 is the shortest.

The 🐟 is the longest.

p53 Answers will vary.

p54

This 🍶 is as tall as 7 🔲 .

This 🌻 is as tall as 4 🔲 .

p55

p56

p57

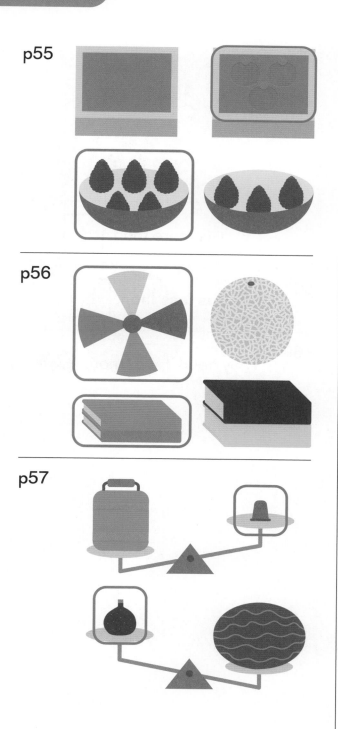

p58

p59

p60

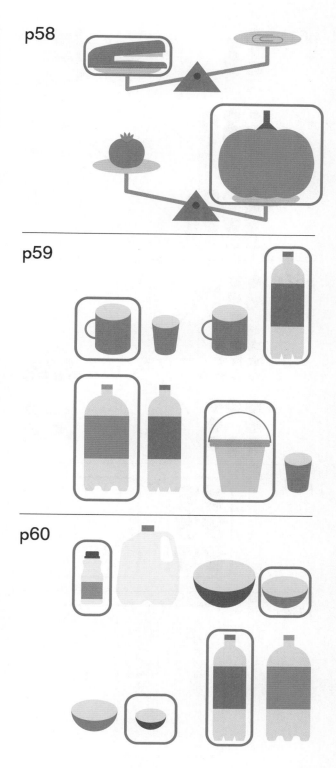

p61

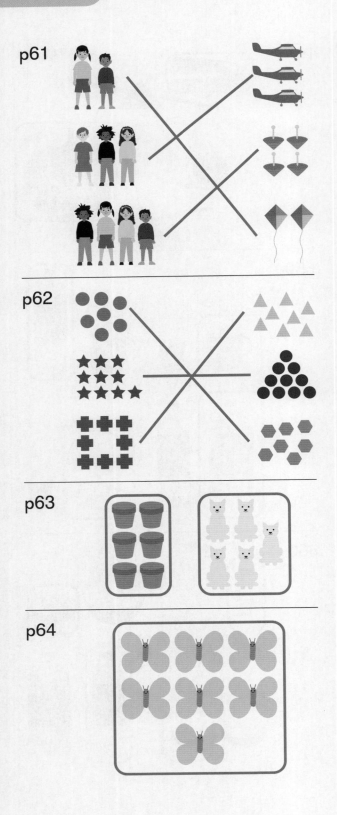

p62

p63

p64

p65

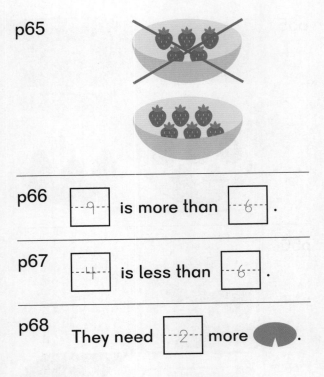

p66 | 9 | is more than | 6 |.

p67 | 4 | is less than | 6 |.

p68 They need | 2 | more ⬤.

Answer Key KB

p69

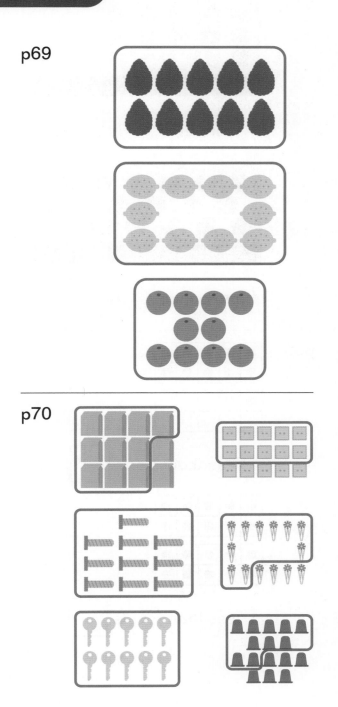

p70

p71

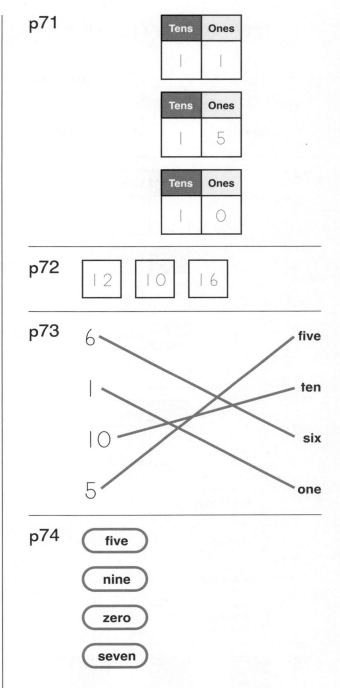

Tens	Ones
1	1

Tens	Ones
1	5

Tens	Ones
1	0

p72

12   10   16

p73

6 —————— five
1 —————— ten
10 ————— six
5 —————— one

p74

five

nine

zero

seven

Test 12

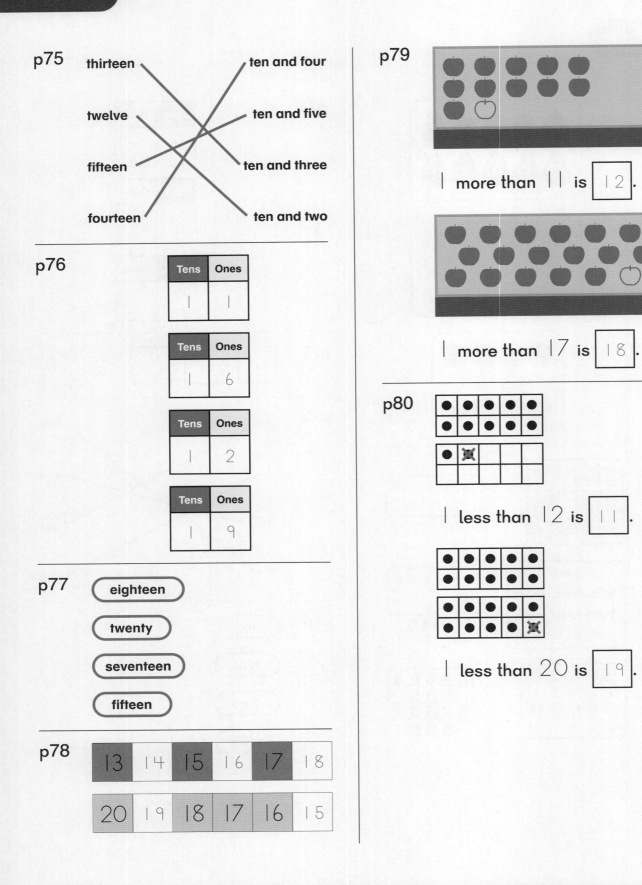

p75

thirteen — ten and three
twelve — ten and four
fifteen — ten and two
fourteen — ten and five

p76

Tens	Ones
1	1

Tens	Ones
1	6

Tens	Ones
1	2

Tens	Ones
1	9

p77

eighteen

twenty

seventeen

fifteen

p78

| 13 | 14 | 15 | 16 | 17 | 18 |

| 20 | 19 | 18 | 17 | 16 | 15 |

p79

1 more than 11 is 12.

1 more than 17 is 18.

p80

1 less than 12 is 11.

1 less than 20 is 19.

p81

| 1 | and | 2 | make | 3 | .

| 3 | and | 2 | make | 5 | .

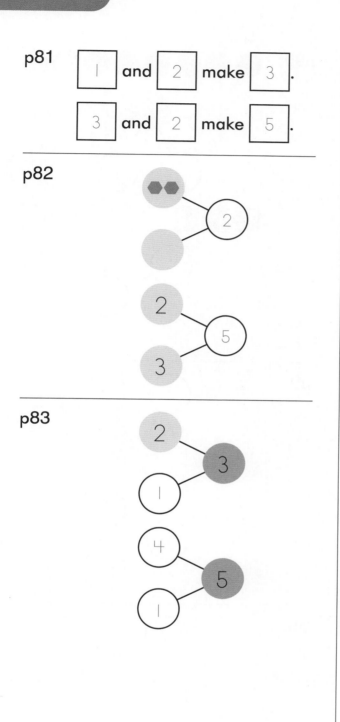

p82

p83

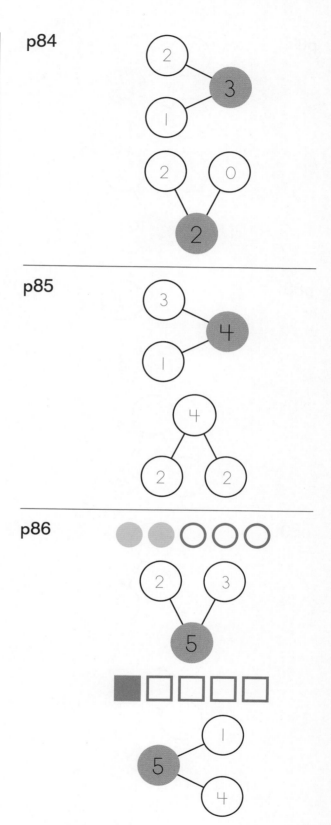

p84

p85

p86

p87

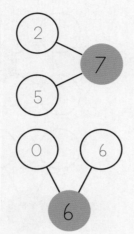

p88

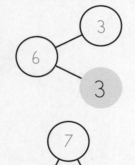

p89

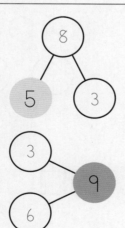

p90

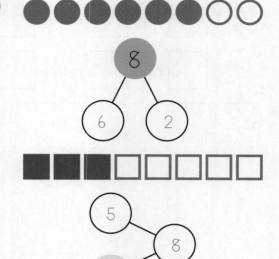

p91

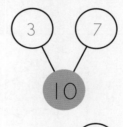

p92

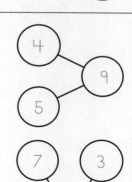

p93 $\boxed{5} = 3 + 2$

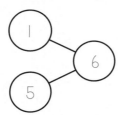

$\boxed{1} + \boxed{5} = \boxed{6}$

p94 $4 + 2 = \boxed{6}$

p95

$5 + 0 = \boxed{5}$

$4 + 3 = \boxed{7}$

p96 $1 + \boxed{4} = \boxed{5}$

$\boxed{3} + 2 = \boxed{5}$

p97 $6 + \boxed{2} = \boxed{8}$

$\boxed{4} + \boxed{3} = \boxed{7}$

p98 Drawings will vary.

$6 + 3 = \boxed{9}$

p99

$3 + \boxed{1} = 4$

$4 = \boxed{4} + 0$

p100

$\boxed{2} + \boxed{3} = 5$

$\boxed{6} = \boxed{3} + 3$

p101

$3 + \boxed{5} = \boxed{8}$

$\boxed{5} + 2 = \boxed{7}$

p102

$10 = \boxed{2} + \boxed{8}$

$\boxed{4} + \boxed{5} = 9$

p103

$\boxed{5} = \boxed{3} + \boxed{8}$

$\boxed{3} + \boxed{7} = \boxed{10}$

p104

$7 + \boxed{2} = 9$

$9 = \boxed{7} + 2$

$4 + 6 = \boxed{10}$

$10 = \boxed{4} + 6$

p105

2 remain.

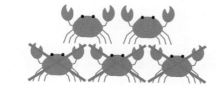

2 remain.

p106 $8 - 2 = \boxed{6}$

$5 - 5 = \boxed{0}$

p107 $7 - \boxed{3} = \boxed{4}$

$\boxed{8} - 1 = \boxed{7}$

p108 $\boxed{9} - \boxed{7} = 2$

$\boxed{6} - \boxed{6} = 0$

p109 $\boxed{5} - \boxed{1} = \boxed{4}$

$\boxed{2} - \boxed{2} = \boxed{0}$

p110

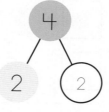

$4 - 2 = \boxed{2}$

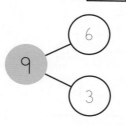

$9 - 3 = \boxed{6}$

p111 4 – 2 = ☐ 2

9 – 3 = ☐ 6

p112 5 – 2 = ☐ 3

5 – ☐ 4 = 1

p113 8 – ☐ 2 = ☐ 6

8 = ☐ 9 – ☐ 1

p114 5 – 1 = ☐ 4

☐ 3 = 7 – 4

p115

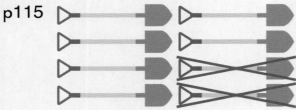

8 – 2 = ☐ 6

10 – 4 = ☐ 6

p116 9 – 5 = ☐ 4

6 – 6 = ☐ 0

p117

$8 + 1 = \boxed{9}$

$9 - 1 = \boxed{8}$

$1 + 8 = \boxed{9}$

$9 - 8 = \boxed{1}$

p118

$5 + 2 = \boxed{7}$

$7 - 2 = \boxed{5}$

$2 + 5 = \boxed{7}$

$7 - 5 = \boxed{2}$

p119

$\boxed{2} \enspace (+) \enspace \boxed{3} = \boxed{5}$

Alex and Emma have

$\boxed{5}$ 🎈 altogether.

p120

$\boxed{7} \enspace (+) \enspace \boxed{2} = \boxed{9}$

There are $\boxed{9}$ 🦊 altogether.

p121

$\boxed{6} \enspace (-) \enspace \boxed{3} = \boxed{3}$

$\boxed{3}$ 🐭 remain.

p122

$\boxed{8} = \boxed{5} \enspace (+) \enspace \boxed{3}$

$\boxed{7} \enspace (-) \enspace \boxed{4} = \boxed{3}$

Test 20

p123

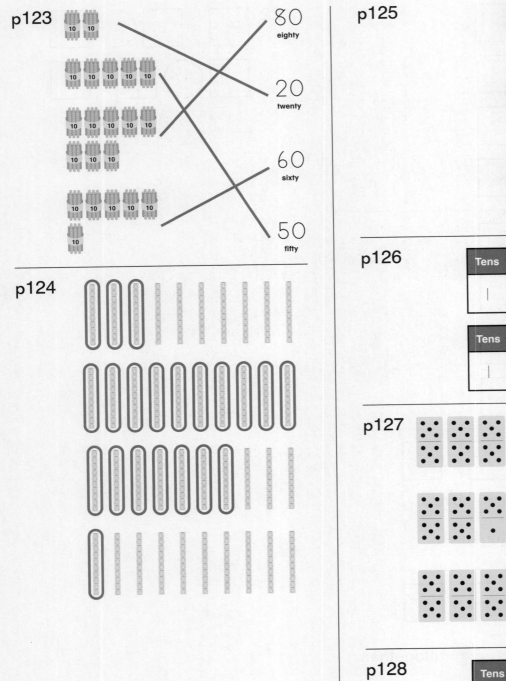

p124

p125

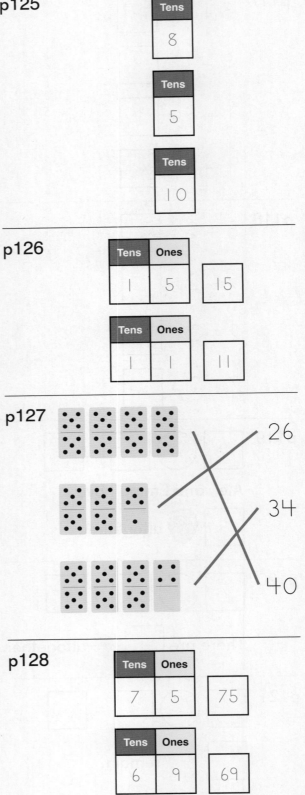

Tens
8

Tens
5

Tens
10

p126

Tens	Ones
1	5

15

Tens	Ones
1	1

11

p127

26

34

40

p128

Tens	Ones
7	5

75

Tens	Ones
6	9

69

p129

Tens	Ones
9	7

97

Tens	Ones
7	9

79

p130

(8 3)

(9 9)

p131

Tens	Ones
7	1

71

Tens	Ones
6	0

60

Tens	Ones
9	5

95

p132 2, 10, 14, 26, 27, 31, 43, 50,
59, 61, 88, 95, 100

p133 35 45

p134 10 20 30 35

p135

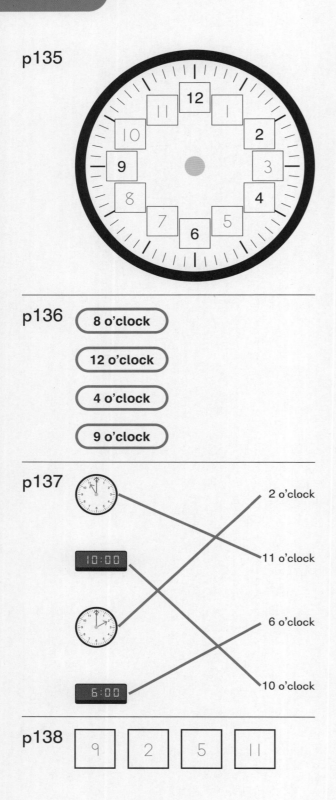

p136

(8 o'clock)

(12 o'clock)

(4 o'clock)

(9 o'clock)

p137

2 o'clock

11 o'clock

6 o'clock

10 o'clock

p138

| 9 | 2 | 5 | 11 |

p139

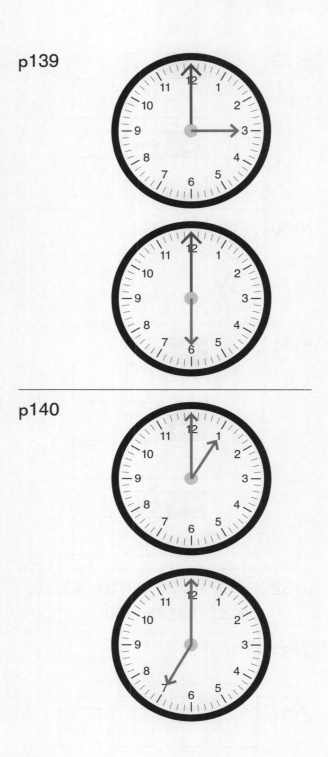

p140

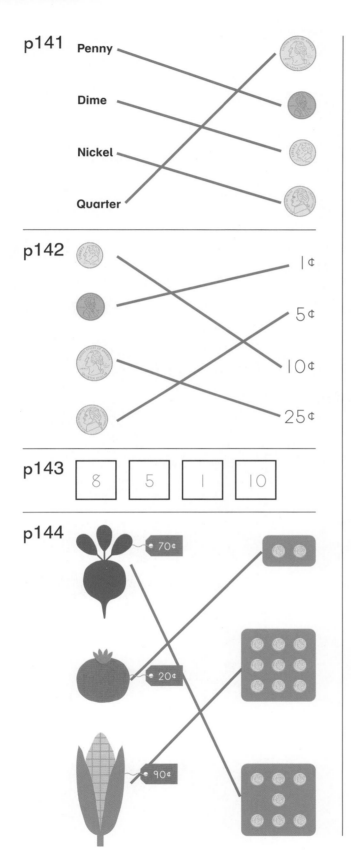

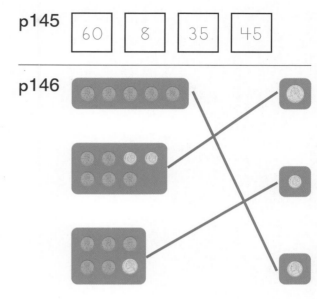

p141

Penny
Dime
Nickel
Quarter

p142

1¢
5¢
10¢
25¢

p143

| 8 | 5 | 1 | 10 |

p144

70¢
20¢
90¢

p145

| 60 | 8 | 35 | 45 |

p146

BLANK